Science and the history and future of humanity

A perspective on civilisation

from the 2nd Law of Thermodynamics

By Martin Gellender

ISBN 978-0-646-81076-8

Science and the history and future of humanity

A perspective on civilisation from the 2nd Law of Thermodynamics

1. Introduction

You would have discovered the 2nd Law of Thermodynamics, whether or not you think you did. By the time you were ten years old, you would have observed and experienced it. You would have noticed that a cold bottle of soft drink, left on the kitchen counter on a summer day, gradually warms to room temperature. You would have noticed that a hot cup of tea gradually cools to room temperature.

You would have noticed that the opposite doesn't happen. When you take a warm bottle of Coke off the grocery shelf on a summer day, it doesn't get cold by itself. Of course, you would know that it is indeed possible to remove heat from a bottle of soft drink and get a refreshing ice-cold drink, but his requires an active process that takes work. In this case, the work is done by an electric motor driving a compressor inside a refrigerator.

If you had thought about it, you would have realised that heat flows from a warm object (like the air in your kitchen) to a cold object (a bottle of soft drink). This process happens spontaneously – all by itself, - without us having to do anything.

This process can be reversed. Heat **can** be transferred from a cold object to a warm object - but only by doing work. That, in a nutshell, is the 2nd Law of Thermodynamics, and it has huge consequences that you probably haven't thought about.

If the 2nd Law of Thermodynamics is universal, applying everywhere at all times – as the evidence suggests it does – then it "sets the ground rules" for everything that happens on Earth and throughout the universe.

We humans are very clever at inventing technology to push the boundaries of what we can do, and what we can't. Any technology that we invent, or might invent in the future, is subject to the Second Law of Thermodynamics. The 2nd Law is our guide to what is possible, and what is not. But, we shouldn't regard the 2nd Law as a sharp boundary for human endeavour. A knowledge of the 2nd Law provides insights into ways to get around its limitations, just like a clever accountant finds ways to avoid paying tax without violating the tax law. The 2nd Law of Thermodynamics provides insights about how humanity has gotten where we are, and how to best face challenges and opportunities for the future.

Before we delve into the 2nd Law of Thermodynamics, we should first review the First Law, which states that energy is neither created nor destroyed. We can convert one form of energy (like the chemical energy in fuel, for example) into another form of energy (mechanical work or heat), but the total amount of energy at the end of the process will be exactly the same as the total amount of energy at the beginning. From an energy viewpoint, you don't get something for nothing.

Not all energy is equally useful. It takes work to pedal a bicycle to high speed, but air resistance and tyre rolling resistance will eventually convert that work into heat, which is dissipated into the surrounding air. The bicycle will come to a stop, all by itself. Since the total amount of energy remains the same, you might think that we could collect heat from the air and convert it back to work. But this is prohibited by the 2nd Law of Thermodynamics (for the same reason that a bottle of soft drink doesn't get cold by itself). Energy is not lost when work is converted into heat, but its ability to do useful work is degraded. We can reverse the process, and get back to the starting point, but only by investing more work.

This book considers how the 2nd Law of Thermodynamics impacted on our human journey to where we are today, and will influence where humanity will go in the future. It puts into perspective how the human population has exploded to take over the Earth, and how the "bronze age" and "iron age" transformed civilisation. It helps us to understand what happens in our own atmosphere, and the atmospheres of other planets. It provides guidance on how technology might address water scarcity affecting many parts of the world, and how food production might be expanded to feed a growing population.

The first four chapters deal with topics that might appear to have nothing to do with the 2nd Law. The early chapters introduce basic concepts that are necessary to understand what the 2nd Law is telling us. I have tried to build up, chapter-by-chapter, the basic knowledge and concepts that I'll draw upon in later chapters to arrive at fundamental laws of nature and, perhaps, at a new way of thinking about the world. In my arguments, I won't ask you, the reader, to take my word for anything. I'll apply basic principles that have been known for decades or centuries. These are principles of which you are probably already aware, and which you have probably experienced in your day-to-day life. Perhaps, my arguments will simply quantify in mathematical form some ideas of which you are already familiar. I know that many people feel intimidated by mathematical equations, bu I ask you to just "go with the flow". Mathematics is a powerful tool that can crystallise underlying patterns of meaning that we would otherwise not see.

So, let's start with where we came from.

2. Exponential growth of populations

All living creatures reproduce to survive. If a species is to maintain its numbers, each generation must have the capacity to produce sufficient offspring to replace itself. In an adverse environment, some offspring will not survive predators, disease or food shortages, so each generation must have the capacity to produce *more* offspring than is needed simply to replace its numbers.

Over the long term, through periods of adversity and abundance, a stable population must be in equilibrium with its environment. Successive generations must produce just enough surviving offspring to replace itself.

In favourable environments, where ample food and resources are available, animals, plants and microbes tend to produce more offspring than is needed to maintain a stable population. If this situation continues for multiple generations, a "population explosion" results, with species reaching plague proportions. For animals with long lifetimes, producing relatively few offspring, many decades may be required for the population to grow. On the other hand, small animals and insects often grow to maturity and reproduce within a few months, producing hundreds of eggs, babies or larvae. Bacteria are even more prolific, and some can reproduce every few hours or days (or even, as I've been informed, within 20 minutes!).

Discovery of a new world

Let's imagine that we are members of a bacteria species which, when provided with adequate food and nutrition, doubles in population each 24 hours. Now, imagine that a gust of wind sweeps you up from a local swamp where you and your ancestors have lived for countless generations. You are blown high into the sky, travel vast distances and suddenly . . . blow through the open window of a biochemistry laboratory at the University of Queensland. You come to rest in a petri dish filled with nutrient solution - an apparently boundless source of food and resources.

You have discovered a new world, uninhabited and rich in resources – perfect for establishing a new civilisation for your species. And this is exactly what you do.

Being asexual, you reproduce into two offspring after one day. On the second day, your direct offspring reproduce, giving rise to four residents of the new community. On day three, the population reaches eight bacteria. By the tenth day, there are 1,024 members in the new colony. In this case, with a "doubling time" of one day, the number of bacteria **N** is related to the time **t** (in days) that have elapsed by the equation:

$$N = 2^t$$

The growth of population during the first week of the fledgling colony is shown in this graph.

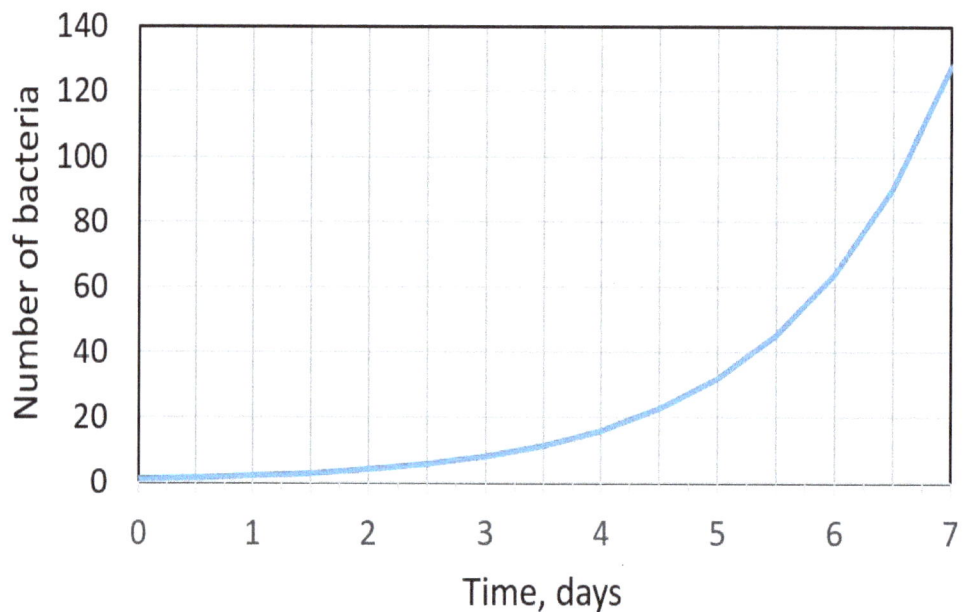

But things don't stop here. There is still plenty of food and resources for the growing community. Twenty days after your arrival in the new world, there are more than a million residents in the community. **By the end of the first month, the population has reached one billion** – about the same as the human population at the onset of the 20th century. The increase in population over time is shown in this graph, with the population scale now shown in millions. Notice that the curve hardly moves off the x axis until about day 24, and then takes off!

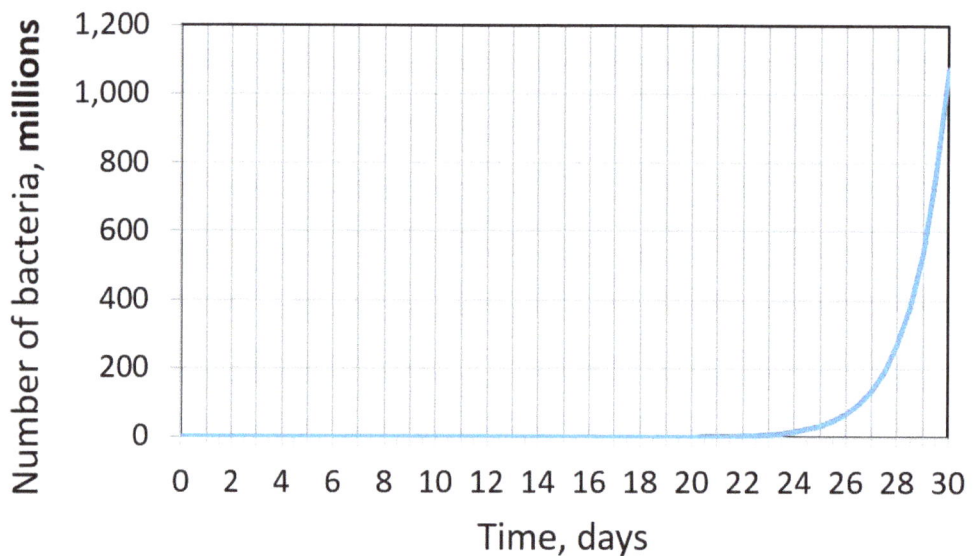

The Chief Bacterium at the Ministry for Health is alarmed. He (or rather, it, since this bacteria is asexual) reports that the food and resources for the colony are running out. Since the first colonizer arrived a month ago, about half the available resources has been consumed. **While half the original resources remain, there is only enough food left for one more day, and then the colony will starve!**

The colony is mobilised into action, sending out adventurous bacteria to explore the surrounding universe. One excited bacterium returns, reporting that another nutrient-rich petri dish is located just alongside the colony. This is an entirely new world, with the same potential for exploitation as the original colony.

A new world is discovered!

Source: https://smart.servier.com/smart_image/petri-dish-2/ Creative commons

Plans are immediately formulated to relocate bacteria from the current overcrowded petri dish to the new colony. However, even if huge numbers of bacterial residents are quickly and efficiently transferred to the new world, this will only extend the colony's food supply **by one day**. Then, the bacterial empire would have exploited all available resources, and the population will collapse.

The moral of this story is that exponential growth cannot continue indefinitely. It eventually reaches a point where the population (or other variable) explodes into huge numbers that can no longer be sustained with finite resources. To avoid catastrophic collapse, growth must stop and be reversed well before the limits of the environment are approached.

You might argue that I have chosen an extreme example of exponential growth; a bacterial species whose population doubles each day. Human populations grow over a much longer time frame. Still, over the past 100 years, the human population has roughly doubled every 40 years. While the rate of population growth has slowed in many countries, the human population will very likely double again during this century (barring a global catastrophe).

Experience over the past few centuries has shown that, as the rate of child mortality and disability has declined markedly, people have fewer children. It seems (and this makes perfect sense to me) that, if parents are confident that they will have 2 or 3 healthy children that will grow to adulthood, they generally won't need or want to have 5 or 6 kids.

In every region of the world, the number of children born to each woman has been reducing in the last few decades. Improving nutrition, education and medical services leads to fewer children being born! Countries which have high rates of vaccination, good nutrition, good sanitation and good medical services are those whose populations are stable or shrinking. Even in countries with continuing high rates of population growth, the number of children per family has declined (in some cases from 6 or 8 per woman down to 3 or 4). So, in the longer term, reduction in poverty and disease actually translates into lower birth rates. The big question for poor countries with high birth rates is: "can they improve nutrition, education and medical services (with accompanying reduction in birth rate) before their expanding population makes this impossible to achieve".

It could be argued that our current situation, as a human species, is similar to the bacterial colony that has consumed half the available resources in its petri dish. Humanity has already depleted, degraded or is already using much of the Earth's resources of ground water, surface water, fish stocks, fertile soils, high-quality petroleum resources, and deposits of phosphate and some metals. One proposed solution is for humans to establish permanent settlements on other planets in our solar system - effectively to colonise another petri dish. Leaving aside the enormous technical challenges and costs, colonising another planet could provide a "lifeboat" for small groups of humans to survive a global catastrophe on Earth, but it would do virtually nothing to alleviate pressures of population growth and depletion of resources on Earth.

Population growth in bacteria and humans is one example of "exponential growth", which arises frequently in the natural world, although – as we have seen - it can continue only over a limited duration, or within a limited range. *"Exponential growth" occurs whenever a population (or any variable) increases at a rate which is proportional to the population.*

Don't think that "exponential growth" only applies to populations of humans, bacteria and other organisms. It arises in an extraordinarily diverse range of situations, which is why I have chosen to discuss it here. Basic knowledge of exponential growth provides us with a valuable tool to understand why our physical world works the way it does.

The "time constant" of exponential growth

Let's consider the population of a small country with one million residents. Of course, the population consists of people of all ages. Each year, babies are born, some people die, some people leave to go elsewhere, and some immigrants arrive from overseas. The net effect is that, each year, the population grows by 1% (a bit less than the rate of population growth in Australia).

Let's consider, what would be the population of this country in 100 years?

At first thought, you might think "the population is increasing at the rate of 10,000 per year (1% of one million), so after 100 years, the population would have doubled to two million. But this is not what would happen.

Why is this? After one year, there will be 1.01 million people living in the country. In the second year, the number of new arrivals will be equal to 1% of the population **at that time** - but now there are slightly more than one million people. The increase in population will be 10,100 – slightly more than the year before. At the end of the second year, the population will be just over 1.02 million, and the population will increase by 10,200.

As the population grows, the **increase** in population each year also grows. The increase in population each year gets larger, as the population gets larger. In fact, the population will have doubled after only 69 years have elapsed.

After one hundred years of 1% population growth, the population of the country will be more than the 2 million that we might have initially expected. In fact, as it turns out, the population would then be 2.718 million.

If we plotted a graph showing how the number of people grows with time, it would be a typical exponential growth curve. Note that, at the initial time zero, the population is growing at the rate of 10,000 per year. **If** the population continued to grow by 10,000 per year (as shown by the dotted red line), it would reach two million after 100 years. But, as the number of people increases, **the rate of growth increases**. After 100 years, the population reaches 2.718 million (as shown by the blue curve).

Exponential growth of a population

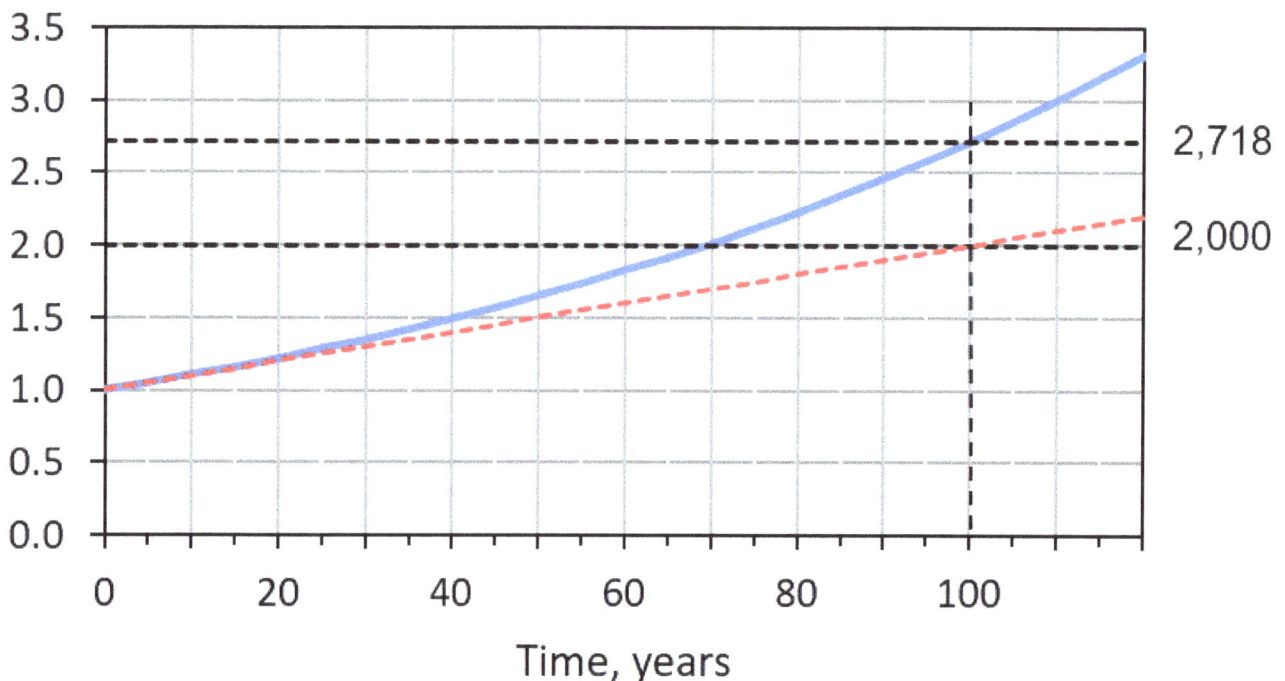

The characteristic time constant is the reciprocal of the rate of increase per unit time. For a population growing at 1% per year, its the characteristic growth time t_o is 100 years. If the initial population is one million people, we will find 2.718 million people after the first hundred years has elapsed. After two centuries, we find 2.718 x 2.718 = 7.38 million people. After 3 centuries, we find 2.78 x 2.718 x 2.718 = 20.1 million people. After each century, the population increases 2.718 times (or roughly, three times).

Scientists use the mathematical notation e^n to show that e is multiplied by itself n times. For example:

$e^0 =$	1.00
$e^1 =$	2.71
$e^2 = 2.718 \times 2.718$	$= 7.38$
$e^3 = 2.718 \times 2.718 \times 2.718$	$= 20.1$
$e^4 = 2.718 \times 2.718 \times 2.718 \times 2.718$	$= 54.5$

In exponential growth, the number N of people, bacteria or whatever is given by the initial population N_o (at time zero) multiplied by the number e multiplied by itself for the number of time periods t/t_o that elapse:

Equation (1)

$$N = N_0\, e^{t/to} \qquad \text{Exponential growth}$$

Where N is the number after time t.
N_o is the original number at time zero.
t is the time that elapses
t_o is the characteristic time during which N increases 2.718 times

The number of people who have ever lived

While the concept of exponential growth of populations is a very useful concept, populations rarely grow at a fixed rate for more than a few generations. The history of humanity is a story of relatively stable periods of growth punctuated by wars, disease outbreaks or drought that wiped out much of the population.

In nature, many species of animals (particularly locusts and rodents) exploit periods of favourable weather and/or ample food sources to undergo a "population explosion", The population undergoes exponential growth until it reaches plague proportions. Eventually, the plague species consumes the available food supply while, at the same time, the population of predators also increases rapidly to feed on the plague species. Consequently, "population explosions" are often followed by episodes of population collapse.

In the same way, human population has probably followed the same pattern of alternating periods of growth and decline. However, throughout much of mankind's existence, humans lived in isolated settlements and communities. While populations in one area might undergo rapid growth or catastrophic decline, this would be offset by different conditions in other geographical areas, causing global population to be more stable. If we average out short-term variations, the human population has undergone steady periods of exponential growth for long periods of time.

Currently, about seven billion people are alive. It is interesting to compare that with the number of humans who have ever lived. That's a tricky question, as we would need to define a starting point. As we go back further and further into pre-history, it becomes increasingly difficult to estimate the human population, and to define which creatures we consider to be "human". Do we include Neanderthals, Denisovans and other extinct hominin species that may, or may not, have been among our ancient ancestors?

Some researchers have been struck by the very limited genetic variation among the current human population. They have suggested that the human population suffered catastrophic decline about 75,000 years ago, perhaps caused by a super-volcano eruption at Lake Toba in Indonesia that would likely have dramatically affected the world's climate for decades. According to this theory, the human population was reduced to perhaps only a few thousand individuals, and the survivors may have been limited to those living within a few coastal areas (with access to seafood resources). Presumably, everyone alive today would be a descendent of these few thousand people. Such a "population bottleneck" at that time would account for the limited genetic diversity of the human population.

If we assume that the "Toba Catastrophe Theory" is correct, then the human population grew from, say, 10,000 individuals at the time of the catastrophe to the current population of 7 billion, an increase of 700,000 times. Let's say that the population grew at a constant exponential rate, with an average characteristic time t_o for the population to grow by a factor of e – that is, by nearly 3 times. We can substitute into Equation (1) that the population grew by a factor of N/N_o = 700,000 over a period t of 75,000 years, and solve for t_o. This gives a characteristic period of about 5,500 years for the population to increase 2.7 times. That seems an extraordinarily long period, particularly in light of the fact that the human population has grown more than threefold within the past 100 years! But bear in mind that the 5,500 years would be the **average** time for the population to grow nearly three times, taking into account numerous episodes of famine, wars and disease, when the population could have undergone sizeable reductions.

Let's accept for now that these assumptions are valid, and look at the long-term growth of the human population (averaging out short periods of rapid growth and catastrophic wars, disease, etc). We consider the period about 75,000 years ago as the beginning (t = 0), when the human population started at 10,000 people. From then, the population N grew exponentially with a characteristic time period of 5,500 years, according to the equation:

$$N = (10,000)\ e^{t/5,500}$$

Consider a graph showing how the population varied over time, starting at 75,000 years ago (with 10,000 people), and extending to the present time (with a human population of around 7 billion). **The area under this graph gives the total number of person-years** during the past 75,000 years. As it turns out, the area under the exponential growth graph is simply given by the current population N (7 billion) times the characteristic time t_o (5,500 years), or about 38.5 trillion person-years.[Note 2]

During the past 75,000 years, humans might have lived an average lifetime of 50 years (life was much shorter in ancient times, and in particular, child mortality rates were much higher). So, by calculating the total number of **person-years** over the past 75,000 years (given by the area under the curve), we are counting each person 50 times. To get the number of people who were alive during the past 75,000 years, we must divide the number of person-years by the average lifetime, giving 770 billion people. In other words, **the current population of the Earth represents less than 1% of the total number of people who have lived during the past 75,000 years**.

As it turns out, we would derive pretty much the same result even if there had been no "population bottleneck" – just so long as the population grew at a steady exponential rate. If the human population was one million people 75,000 years ago, then the characteristic time period t_o would be

Human population over past 75,000 years

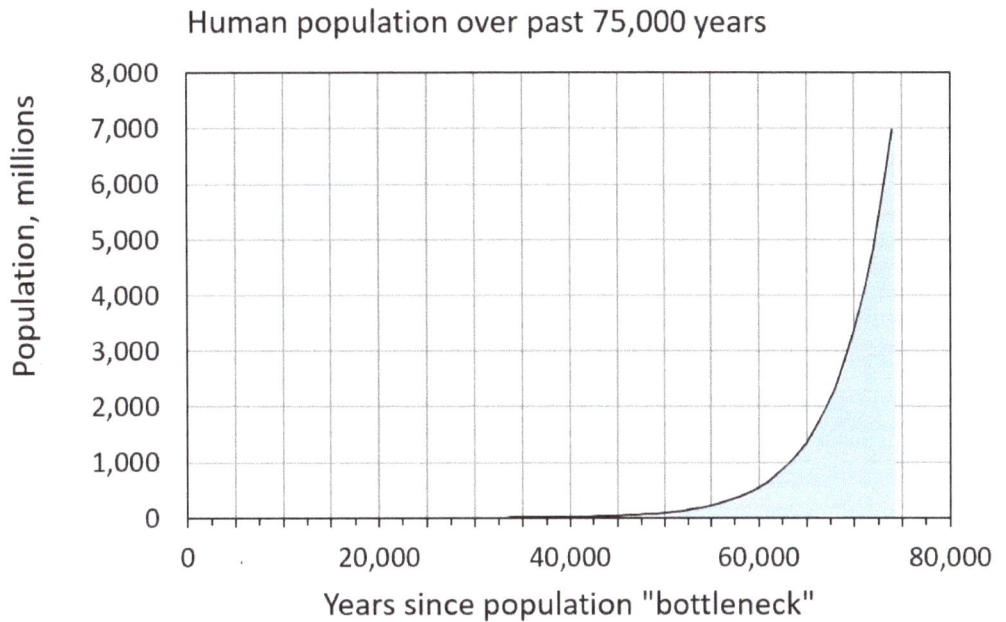

8,500 years, and the current human population would be 0.6% of the number of people who lived during the past 75,000 years.

In fact, we would derive very nearly the same result – that 770 billion people have been alive – even if we extend the starting point back to 200,000 years ago, when the first anatomically-modern humans appear in the fossil record.[Note 3]

The population of aboriginal Australians

Consider the arrival of Aborigines in Australia, which probably occurred about 60,000 years ago. The "time zero" of human settlement on the Australian continent may have occurred as a single arrival of a small community (of, say, perhaps 100 people). As the population became established and grew, they could expand across the entire continent of Australia until being limited by the food resources available by hunting and gathering native flora and fauna. At the time of European settlement, the aboriginal population of Australia is estimated to have been about 750,000.

Let's assume that the Aboriginal population of Australia grew at a constant rate throughout the 60,000–or so years since "time zero" of their arrival. If we assume that the population grew from an initial settlement of 100 people to 750,000, then Equation (1) allows us to determine the characteristic time constant (for the population to grow 2.718 times) as 6,700 years. This is a growth rate of only 0.015% - some 100 times less than the current population growth of Australia!

If we plotted a graph of the number of aboriginal Australians versus time, with a characteristic growth rate of 6,700 years, then the area under the curve is equal to the number of aboriginal person-years in Australia over the past 60,000 years. If the average lifespan of aboriginal Australians during this period was 50 years, then **the total number of aborigines who have lived in Australia is more than a hundred times (6,700/50) greater than the number living at the time of European settlement.**

During the 60,000 years that aboriginal people lived in Australia, their hunter-gather lifestyle would probably have changed relatively little, and would have been familiar to those living at the time of European settlement. Perhaps then, it should be no surprise that ancestors play a prominent role in aboriginal culture. This would likely create a mindset that each person's life, and the lives of others in their generation, is a momentary interlude along a vast continuum of human experience.

Compare this with modern society, which hardly resembles our ancestor's life prior to the industrial revolution. If those of us of European descent could meet our ancestors from before the industrial revolution, we would have as little understanding of their way of life as they could understand our internet-connected world. Our daily lives in 21st century Australia have virtually nothing in common with our distant ancestors living as peasants (or possibly nobility) in 15th century feudal Europe. We could probably only understand and empathise with our ancestors who were alive during the past 200 years, who might share some of the same problems, achievements and issues facing us today.

During the past two centuries, rates of population growth have averaged about 1-2% per year. The characteristic growth time (for the population to increase 2.7 times) has been of the same order as the average human lifespan. As a result, the number of people alive in the world today is roughly comparable with the total number of our ancestors who were alive during the past 200 years. We have little or no shared cultural connection with 99% of our ancestors who lived during the past 75,000 years. Perhaps this contributes to a lack of interest in our ancestral origins and a narrow focus on ourselves and the present time.

The view from the opposite perspective

We can also look at exponential growth from the opposite perspective. Instead of considering how the number **N** of bacteria or people increases with time, let's consider the time **t** required for the number of bacteria to reach a certain level.

Consider a bacteria species which multiplies exponentially with a time period of one day. At time zero, we have 1,000 bacteria. We must wait one day for the initial population to increase 2.718 times. The population grows 2.718 times again on the second day, and then another 2.718 times on the third day. If we plot a graph of the time elapsed versus the number of bacteria it looks like this:

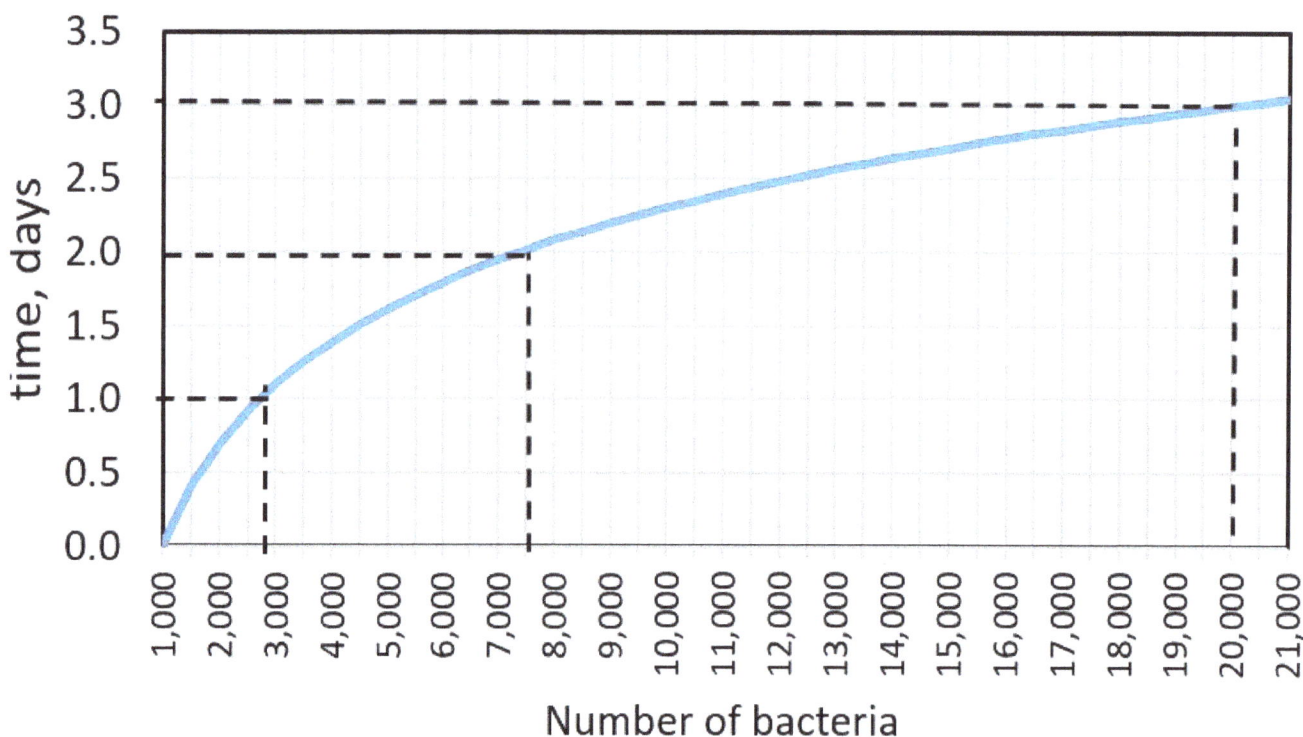

The relationship can be expressed in a mathematical equation as:

Equation (2) $\qquad$ Time $t = t_0 \ln\left[\dfrac{N}{N_0}\right]$

Where **t** is the time that elapses
t_0 is the characteristic time (for the population to increase 2.718 times)
N is the number of bacteria after time **t**
N_0 is the initial number of bacteria at the initial time $t = 0$

N/N_0 is simply the ratio of the final population relative to the initial population. The term **$\ln(N/N_0)$ *is the number of times e must be multiplied by itself to give the number* N/N_0** . So, for example:

$\ln(1)$ $\qquad$ = 0

$\ln(2.7)$ $\qquad$ = 1

$\ln(7.3)$ $\qquad$ = 2

$\ln(20)$ $\qquad$ = 3

$\ln(54)$ $\qquad$ = 4

We have seen that exponential terms tend to "explode" into huge numbers (since they increase at an increasing rate). On the other hand, logarithmic terms tend to "plateau" (since they grow at a decreasing rate).

The triumph, and ultimate demise, of "Moore's Law"

One striking example of exponential growth within our lifetimes is the rate of improvement in microelectronics and computer technology. In the early 1960's, electronics manufacturers began to produce "integrated circuits", containing 10 or 20 transistors on a single wafer of silicon. Each transistor is a simple electronic switch, and a few dozen can be connected into circuits that act as amplifiers, timers or other such functions. Soon, manufacturers were producing sophisticated logic circuits (the basis of computers) containing more and more transistors. In 1965, Gordon Moore (co-founder of Intel Corporation) famously predicted that the number of transistors that could be produced on a single chip (and the processing power of the chips) would double every 18 months or so. This rate of growth corresponds to a characteristic growth time of 2.2 years (for the number of transistors to increase 2.718 times).

This prediction, called "Moore's Law", must have seemed totally implausible at the time, but turned out to be perhaps the most prescient and accurate prediction in history. The number of transistors per computer chip, and their processing power, has grown exponentially - exactly as Gordon Moore predicted! Now, 50 years later, each mobile phone contains about **5 billion** transistors, and has far greater processing power than was available to NASA for the Apollo missions to the moon!

Many technologists wonder how long this exponential growth in processing power could continue. Some believe that the semiconductor industry is reaching the "end of the road" for Moore's Law, but others expect the exponential growth in computer processing power to continue for the foreseeable future. Having more and more transistors on a single chip means that the transistors have been getting smaller and smaller. Currently, transistors in integrated circuit chips may contain only a few thousand atoms. It is hard to imagine how such intricate structures can be fabricated on this miniscule scale - let along be made much smaller. If Moore's Law were to continue for another 40 years, a single phone or computer chip would have more processing capacity than all the computers, phones, tablets and laptops that are currently operating in Australia. It is inconceivable what astounding, frivolous or malign functions could be done with such processing power.

About twenty five years ago, I attended a renewable energy conference and heard a plenary lecture presented by Dr Andrew Bartlett on the topic of exponential growth. Dr Bartlett was the oldest presenter at the conference, by far, and the only one to use an outdated overhead projector with hand-drawn slides on plastic sheets. I regret to admit this now, but I initially thought he would be an old "fuddy duddy". As it turned out, he gave the most entertaining, hilarious and informative lecture of the entire conference. You can view a full lecture (about 1-1/4 hours) presented by Dr Bartlett on this topic at: https://youtu.be/sl1C9Dyli_8

As we have seen, exponential growth occurs whenever the rate of growth of a population (or anything else) increases at a rate which is proportional to the population. However, ***there are many situations where a population (or something else) <u>reduces</u> at a rate which is proportional to the population***. This situation, which is called "exponential decay", is probably even more important, and plays a larger role in shaping the natural world. And, as we'll see in the next chapter, exponential decay is relevant to our society, our economy and how we live our lives.

Notes & References
1. Like π, **e** can only be exactly described with an infinite number of digits. A more accurate figure is 2.71828182845, but for all practical purposes, 2.718 will do fine.

2. The equation for exponential population growth is $N = N_o e^{-t/to}$. (where N_o is the initial population at time zero, **N** is the population after time **t**, and **to** is the characteristic time period for the population to grow **e** times). The number of person-years can be found by "integrating" the equation, to get the area under the graph of Population **N** versus time **t**. This gives: **Person-years = $(N - N_o) t_o$**. Normally, we are looking at situations where the current population **N** is much, much larger than the initial population. Then, we can simplify this to: **Person-years = $N t_o$**. To find the total number of people who have lived, we need to divide the number of person-years $N t_o$ by the average lifetime t_{av}.

   Total people who have lived = $N t_o / t_{av}$

3. Humans have been evolving from our primate ancestors for over a million years, generally becoming more and more like us in appearance and mental ability. So where do we draw the line in saying that a particular ancestor is a "modern human"?. A useful guide, in my view, is to ask whether such a person – given a bath, haircut, shave and modern clothes – would raise alarm among shoppers walking through the Queen Street Mall in Brisbane.

3. Exponential decay

Exponential growth is not limited to populations of human societies, bacteria or insects: it applies to a wide range of situations. One situation that is directly relevant to many of us is the growth, or decline, of our financial savings after we retire.

You have probably heard of "the miracle of compound interest", where interest-bearing bank accounts increase greatly in value over an extended period of time. This should be called "the miracle of exponential growth".

Consider a young worker (let's call her Anne) who, ten years ago, had more income than she needed to meet her expenses at that time. She decides to make a one-off investment of $10,000 into a superannuation account. The account would earn dividends, at a rate depending upon general economic conditions, the choice of superannuation fund, and the type of investment mix. For example, the average rate-of-return for the past 10 years of a Qsuper Balanced superannuation account has been 10% per year. Each year, on average, the amount in such an account would grow by one-tenth its value. So, an initial $10,000 investment (at time zero) would have grown to $11,000 after one year. During the second year, and the years after, dividends would be paid on the initial investment of $10,000 – as well as on dividends previously paid. Consequently, after ten years, the fund would have grown **e** times, to $27,180.

If the fund continued to grow at 10% per year (although, this might be unlikely in the current era of low interest rates), it would increase again by a factor of **e** to $73,800 (at time t = 20 years), to $200,100 in another ten years (t = 30 years), and to $545,000 some 40 years after the initial investment. This is fifty times more than the initial investment!

In reality, the growth in value wouldn't be quite as miraculous as this suggests. Firstly fees and taxes would likely apply, reducing the effective rate of return. Secondly, although the number of dollars in the account has grown 54 times in 40 years, inflation would have eroded its value. At the current rate of inflation of about 3%, the amount of goods and services that could be purchased with each dollar would have reduced to only one-quarter of its value 40 years before. Consequently, to assess the growth of purchasing power by a superannuation account or bank account, we should use the "real interest rate" (the rate of interest or dividends *minus* the rate of inflation).

Taking these factors into account, the real rate of growth for a QSuper Balanced superannuation account has averaged about 6% over the last ten years. If this rate of return continues into the future, the purchasing power of an initial $10,000 investment would grow 11 times over 40 years.

Forty years after our young worker has opened the superannuation account, Anne decides to retire. At this point in her life journey, she will no longer be investing money into her superannuation accounts, but will be withdrawing funds to meet her living expenses. By government regulation, retirees are required to withdraw a minimum percentage from their superannuation accounts, depending on age (5% for those aged 65-74), although they can withdraw funds at a faster rate if they wish. Whether or not a retiree can maintain the lifestyle that they want with an income of 5% of their superannuation savings depends on the amount of their savings. If their superannuation account has $500,000, a 5% payment equates to $25,000 per year. For a superannuation account of $1,000,000, payment of 5% equates to $50,000 per year. Thus, someone with limited superannuation savings would likely need to withdraw funds at more than the minimum 5% rate to meet their expenses.

If the rate of dividend payments exceeds the rate of withdrawals, then the account will continue to increase in value. *In this case, the superannuation account will undergo exponential growth at a rate equal to the rate at which dividends are paid into the account minus the rate of withdrawals*.

On the other hand – and this is more likely for many retirees – the rate at which dividends are paid is less than the rate of withdrawals. **In this case, the value of the superannuation account will undergo "exponential decay" at a rate equal to the rate of withdrawals minus the rate of dividend payments**.

The value of a superannuation account will grow or shrink as follows:

Overall Rate of growth (or decrease) =   Rate of dividend payments

– Rate of withdrawals

The rate of growth (or decrease) can be positive (giving exponential growth) or negative (giving exponential decay).

The value (or "principal") of the account **P** will grow or shrink exponentially with time s follows:

Equation (1) $$P = P_o\, e^{(I-w)t}$$

Where P_o is the initial value of the account (at time $t = 0$)
P is the value of the account after time t has elapsed
I is the rate of dividend payments or interest **paid to the account** (as a fraction of the value of the fund)
w is the percentage (fraction) of funds that are **withdrawn from the account** each year
t is the time that elapses

So, if Anne's account accrues dividends at a real rate of 6%, and if she is able to maintain her life-style by withdrawing the minimum 5% of the funds, the account will continue to grow in value at a real rate of 1% per year. But, if the amount in the account is too small for Anne to meet her expenses, she might need to withdraw funds at the rate of, say, 10% per year. Then, the account will shrink by 4% per year.

Here is a graph that I plotted depicting these two scenarios. In one case (orange curve), Anne has accumulated $1,000,000 in her superannuation account during her working life, and is able to maintain her life-style by withdrawing 5% of her account per year. If the account achieves dividend payments of 6% each year, the account increases in value each year. The fund provides Anne with an initial income of $50,000 per year, and her income increases over time.

In the second case (blue curve), Anne has only accumulated $500,000. To achieve **the same initial income of $50,000**, she must withdraw funds at a rate of 10% per year. **The amount withdrawn from the account (10% of the fund each year) is more than the interest or dividents paid into the account (6%).** The amount of her superannuation account undergoes exponential decay, reducing to **1/e** (that is, about one-third) of its initial value in 25 years.

Value of account, funds withdrawal at fixed % real rate of interest of 6%

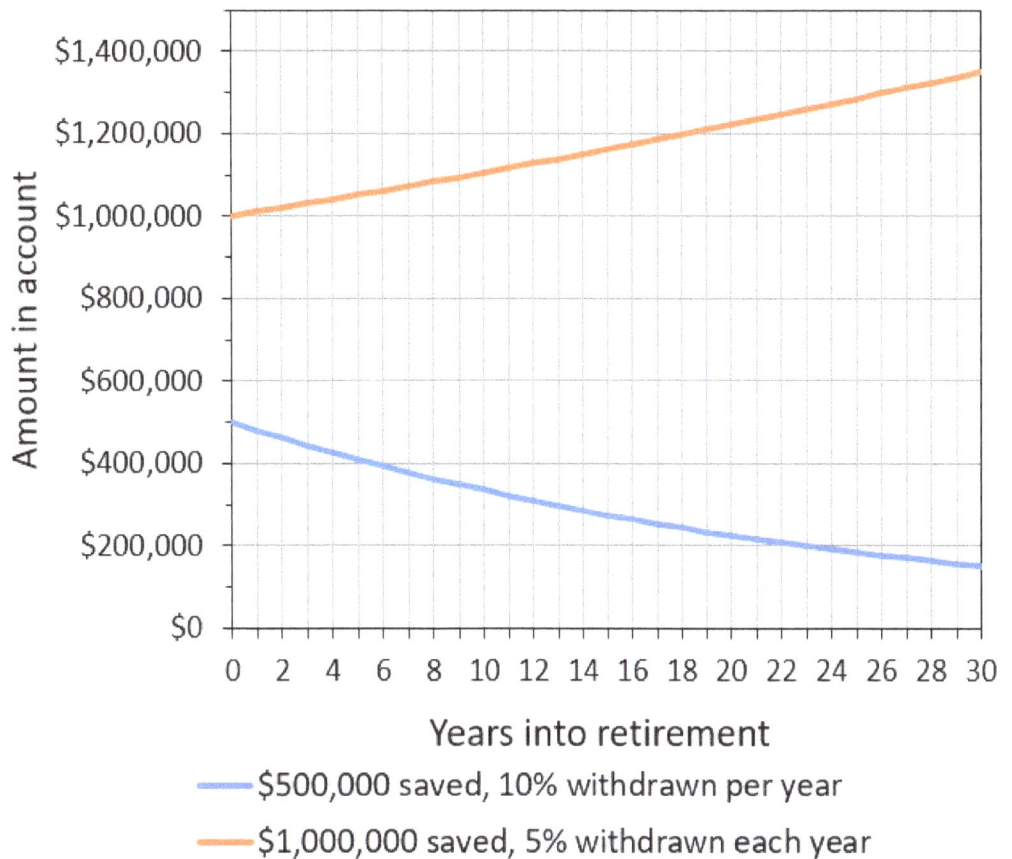

— $500,000 saved, 10% withdrawn per year
— $1,000,000 saved, 5% withdrawn each year

In this case, the state of Anne's finances will deteriorate even more than this graph suggests. While the fund provides Anne with the same initial income of $50,000 per year, the payments reduce over time. As her savings declines, Anne would likely find that she needs to withdraw more than 10% of the fund per year to meet her expenses.

A more realistic scenario for retirees is that, instead of withdrawing funds at a **constant percentage** of the account each year (say, 5% per year), they withdraw a **constant amount** of funds (say, $50,000) per year. The value of the account **P** will then vary over time **t** as follows:

Equation (2) $P = \dfrac{W}{I} - \left[\dfrac{W}{I} - P_0\right] e^{It}$

Where P_0 is the initial value of the account (at time t = 0)
P is the value of the account after time t has elapsed
I is the rate of dividend payments or interest **paid to the account** (as a fraction of the value of the fund)
W is the **amount withdrawn from the account per year**
t is the time that elapses

What does this equation look like in practice? Here is another graph showing the amount in Anne's account for the same two scenarios, but now **with a constant $50,000 withdrawn from the account each year**. Once again, we assume a real rate of dividend payments of 6% per year. The two scenarios differ only in the initial amount of savings ($1,000,000 versus

15

$500,000). In the case of the $1 million superannuation fund (orange curve), the account continues to grow, leaving a sizeable sum to be inherited by Anne's descendants. In the second case (blue curve), Anne's savings will be fully spent after 15 years [Note 1], and she will need to rely on a government pension, and/or support from her family, or sell her house or other assets.

Value of account, fixed $50,000/year withdrawn
real rate of interest of 6%

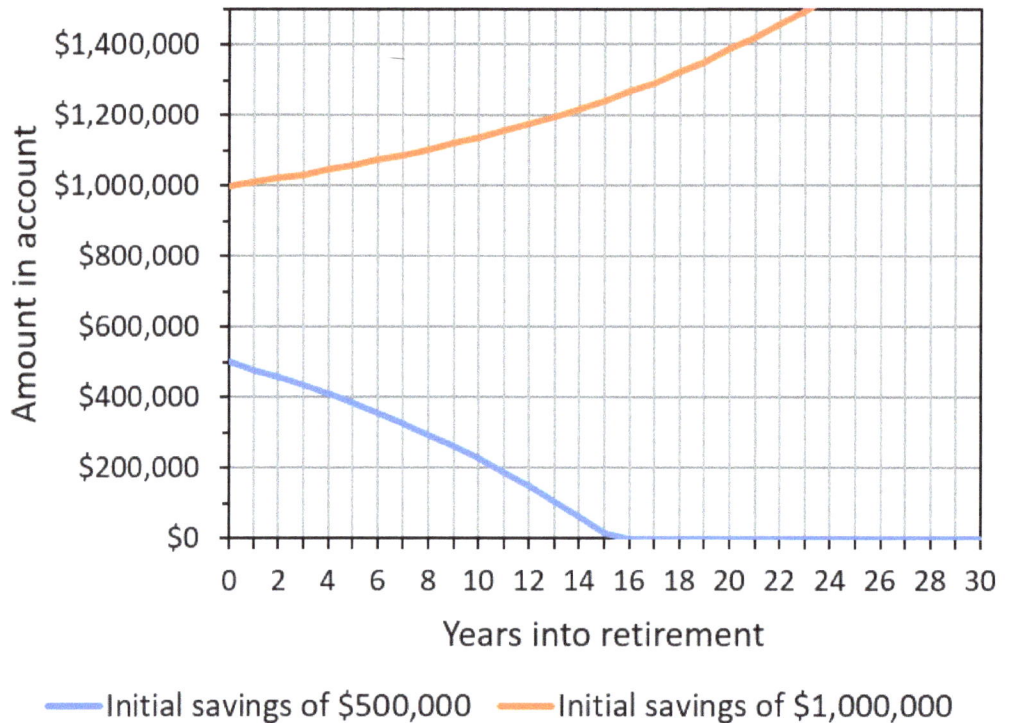

Y-axis: Amount in account ($0, $200,000, $400,000, $600,000, $800,000, $1,000,000, $1,200,000, $1,400,000)
X-axis: Years into retirement (0, 2, 4, 6, 8, 10, 12, 14, 16, 18, 20, 22, 24, 26, 28, 30)

—Initial savings of $500,000   —Initial savings of $1,000,000

Exponential decay and a race for survival

Let's consider a manned spacecraft travelling through the vacuum of outer space. Inside, the astronauts are living within a cacoon that shields them from the deadly conditions just outside their spacecraft. The environment within the spacecraft is at a pressure of one atmosphere, the same pressure which we experience living at the surface of the Earth near sea level. Outside, the pressure is zero.

Suddenly, a leak develops in the outer shell of the spacecraft. Perhaps the spacecraft is punctured by a micrometeorite, leaving a small hole through which air can escape. Or, perhaps a mechanical fault causes air to leak out (as actually occurred in a Russian spacecraft [Note 2]).

Air leak

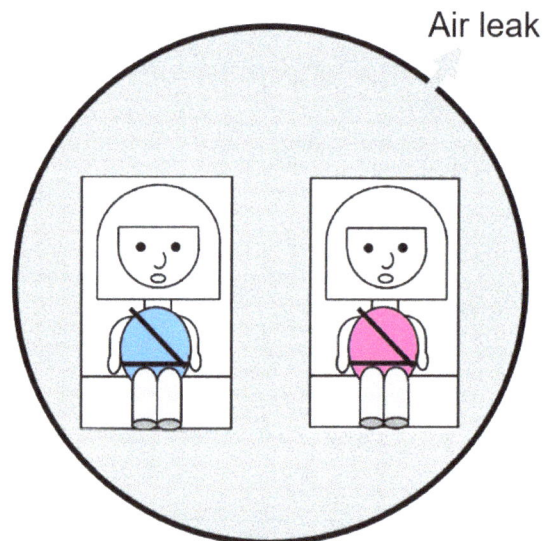

Air begins to escape from the spacecraft, and the pressure inside begins to drop. Let's say that the astronauts check the air pressure gauge and find that air pressure is falling at 1% per minute. How long do they have to locate and fix the leak before the pressure drops so low that they lose consciousness?

When the puncture first occurs (at time t = 0), the internal air pressure is exactly 1.00 atmosphere. After one minute, the pressure has fallen by 1% to 0.99 atmospheres. We might initially think that, with air being lost at this rate, that the pressure would drop to zero after 100 minutes. This pressure loss is depicted by the dotted red line in a graph of air pressure versus time.

But this is not what happens! The situation is not quite so dire. As the internal air pressure drops, the gas escapes at a slower rate. After about 10 minutes, when the air pressure is roughly 90% of its initial value, the rate of loss of air will be 90% of its initial value. As it turns out, after 100 minutes has elapsed, the air pressure will have fallen to 37%, or **1/e**, of its initial value.

Loss of pressure from spacecraft

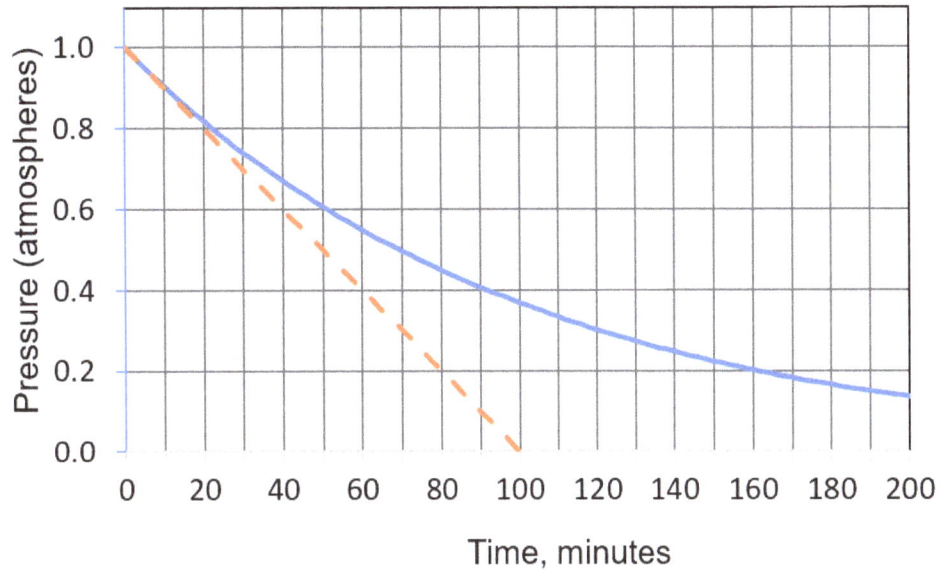

It's critical that the astronauts repair the leak by then, as 0.37 atmospheres is roughly the air pressure at which the astronauts will lose consciousness.

The loss of pressure follows the blue curved line on the graph. This curve is an "exponential decay".

If the leak is still not fixed, the pressure will continue to fall. After 200 minutes, the pressure will fall to 0.37 X 0.37 = 13.5% of its initial value. After 300 minutes, the pressure would be 0.37 X 0.37 X 0.37 = 5% of its initial value. The "characteristic time constant" for the pressure loss is 100 minutes. For each 100 minutes that passes, the pressure falls by a factor of 1/e.

The situation of exponential decay occurs because *the loss of gas is proportional to the amount of gas* in the spacecraft. Exponential decay occurs whenever the loss of any quantity is proportional to its amount at that time. In the case of the leaking spacecraft, let's say that the initial number of gas molecules in the spacecraft (at t = 0, when the micrometeorite hits) is N_o. Some time **t** later, the number of gas molecules has fallen to **N**, where:

Equation (3A) $N = \dfrac{N_0}{e^{t/to}}$ Exponential decay

Where **N** is the number of air molecules after time **t**.
N_o is the original number of air molecules at time zero.
t is the time that elapses
t_o is the characteristic time (100 minutes in this case) during which N reduces to 37% (1/e) of its former value.

The equation is also commonly written in the following form, with a negative sign in the exponent.

Equation (3B) $N = N_o\, e^{-t/to}$ Exponential decay

Note that the "exponent" t/t_o is the number of characteristic time periods that elapse.

The hot coffee dilemma

I enjoyed studying chemistry in high school, and was a good student. I was given a fair amount of homework, but I generally found this challenging and enjoyable. Part of the attraction of sitting at my desk in the evening doing homework, I suspect, was the freezing cold conditions outside on mid-winter evenings. By the time my family finished dinner, it would be well and truly dark outside. Often, the temperature was well below freezing and the wind would be howling outside. On really cold nights, the moisture in the air (from my breath or my mother cooking dinner in the kitchen) would condense as ice crystals on the inside of the windows. I was not distracted by the thought of "going outside to play". In fact, the thought of going anywhere outside our apartment was very unattractive.

I would begin my homework by going into the kitchen to make myself a cup of coffee (instant of course). Nurturing a cup of hot coffee within my hands, as I sat at my desk admiring the intricate ice crystals on my bedroom window, was bliss. Just the warmth in my hands, and the wisp of steam rising from the cup, put me in a state of deep relaxation.

Of course, the coffee would gradually cool and begin to lose its attractiveness. I knew from my physics class that a cup of hot coffee loses heat to its surrounding environment in three ways:

1. Heat passes through the walls of the cup and is absorbed by the surrounding air.

2. The hot outside walls of the cup radiate infrared radiation to the surrounding environment.

3. Heat is lost as water vapour evaporates from the hot coffee.

In each of these cases, *the rate of heat loss is proportional to the temperature difference between the coffee cup and its surrounding environment*. Consequently, the escape of heat from the cup and its reduction in temperature meets the conditions for exponential decay.

The rate of cooling is proportional to the temperature difference between the hot coffee and the surrounding air. The temperature falls rapidly at first, but then - as the cup cools - the rate of cooling gets less and less.

Let's say that the coffee is initially at the boiling point of water, 100°C, and that the surrounding air temperature is 20°C. So, initially, there is an 80 degree temperature difference between the cup and its surroundings. We might find that, after one minute, the temperature has fallen ten degrees. If the temperature continued to fall at this rate (as shown

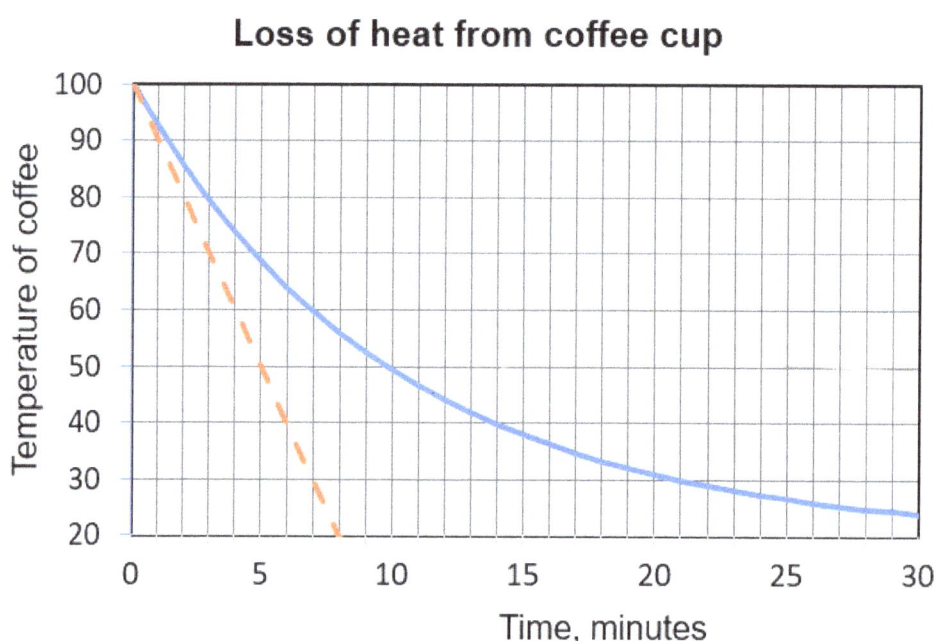

Loss of heat from coffee cup

by the dotted red line on the graph), the coffee would cool to room temperature in a mere 8 minutes. But, as the coffee cup cools, it loses heat as a slower rate, and the temperature will undergo an exponential decay, as shown in the blue curve on the graph. We would find that, after 8 minutes, the temperature difference between the cup and its environment has fallen to 37% (1/e) of its initial value.

One day, just as I was pouring a cup of coffee before starting my homework, I was distracted by a telephone call. By the time I finished the call, my coffee had lost some of its precious heat. This raised a critical dilemma which I have pondered at some length: Would it have been better to add milk straight away, or to add milk after the telephone call?" Since then, I discovered that several other science nerds like myself had considered this very question.

Of course, adding milk cools the coffee, so the immediate inclination might be to hold off adding the milk until you are ready to drink the coffee. But, by adding the milk straight away, you reduce the temperature of the coffee, so less heat loss occurs during the period of the telephone conversation. However, my recent calculations show that the strategy of adding milk immediately has only marginal benefit, causing the coffee to be about 1.5 degrees warmer than it would be if you added the milk later. But even this marginal benefit is uncertain, as other complicating factors are at play [Note 3].

Light intensity in the ocean

Imagine that you are scuba diving in the ocean, or a slightly murky river or lake. Sunlight penetrates the surface of the water, and there is ample light at shallow depths. However, since light is absorbed or scattered by particles suspended or dissolved in the water, the light level gets dimmer and dimmer at increasing depth. Below about 150 metres in the ocean, there is virtually no daylight, and darkness prevails. Many fish species that remain below such depths are blind and rely on other senses to find food and avoid prey: having vision is no benefit if there is no light to see.

Since light is absorbed or scattered by particles of sediment, you might find – in one particular lake, for example - that the light intensity reduces 10% for each metre that you descend. Once again, you might naively expect that the light intensity would reduce linearly with depth, with zero light intensity at 10 metres. But, by now you would know that this is not what we find.

At one metre below the surface, the light intensity is 90% of that at the surface. As the light passes through the second metre of depth, the reduction in light intensity is only 90% as much as occurred in the first metre. At a depth of 10 metres, the light intensity will have fallen to 37% (1/e) the intensity at the surface. At 20 metres in depth, the light intensity reduces to 0.37 X 0.37 = 13.5% of the intensity at the surface.

At 70 metres depth, the light intensity is one-thousandth the intensity at the surface, and at 140 metres depth, the intensity is one-millionth the surface intensity. As we go deeper and deeper, the light intensity gets less and less, but never reaches zero (although the light becomes so dim that is undetectable). At 210 metres, the light intensity is one-billionth as much as at the surface, and at 280 metres, the light intensity is one-trillionth!

Variation of light intensity with depth

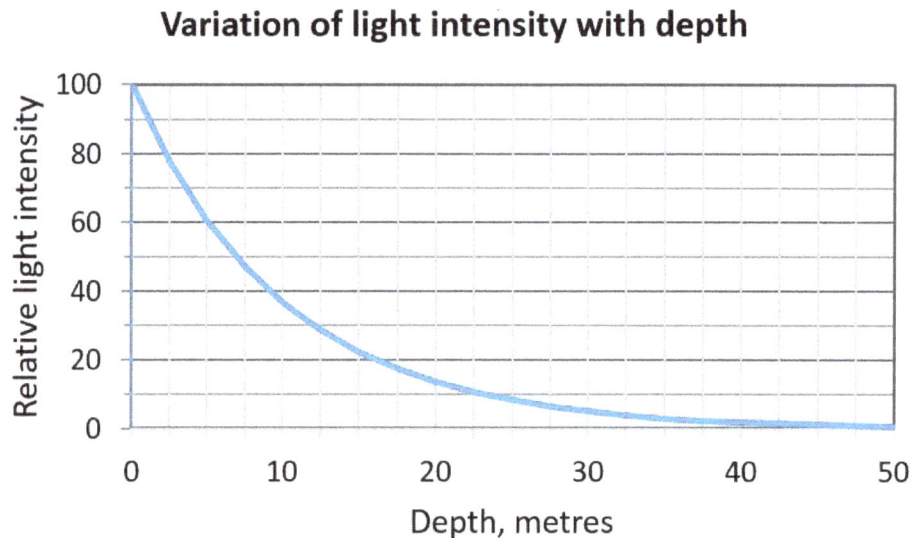

In this case, the light intensity reduces to 1/e of its previous value for each additional 10 metres in depth. If the initial light intensity at the surface is I_o, the light intensity at depth D is given by:

$$\text{Light intensity} = I_o \, e^{-D/10}$$

Exponential decay applies to a wide range of situations. A really important illustration of exponential decay is the reduction in pressure as we rise through the atmosphere. This literally defines and shapes our world, as we live within a remarkably thin film of gas covering the Earth's surface. And, as we'll see in the next two chapters, the exponential decay of atmospheric pressure defines other worlds as well - other planets and moons which have an atmosphere in our solar system and beyond.

<u>Notes</u>

1. The funds in an account will eventually decline to zero if the amount withdrawn each year from the fund exceeds the dividends paid into the fund. How long this will take can be calculated from Equation (2), setting the amount **N** in the account to zero, and re-arranging to find the value of time **t**.

For an initial account balance of P_o, a fixed rate of interest **I**, and a fixed amount **W** withdrawn each year, the number of years **t** after which funds are exhausted is given by:

$$\text{Number of years for funds to be exhausted, } t = \frac{1}{I} \ln \left[\frac{W}{W - P_o I} \right]$$

2. The danger to astronauts of an air leak into space is not hypothetical. On June 30, 1971, three astronauts aboard a Russian spacecraft, Soyuz 11, were preparing to return to the surface of the Earth. Vladislav Volkov, Georgi Dobrovolsky and Viktor Patsayev had just completed a successful mission on the world's first space station. They had spent 22 days in low Earth orbit, a few hundred kilometers above the Earth's surface,

At the beginning of their descent towards the Earth, a faulty valve suddenly opened, allowing air inside their spacecraft to leak into space. The astronauts desperately tried to locate the leak, but were unable to do so. The spacecraft followed its programmed re-entry path through the atmosphere and parachuted to the surface. A waiting ground crew quickly reached the spacecraft and opened the hatch, expecting to congratulate the astronauts on their successful mission, but the three crew were already dead from depressurization of the spacecraft.

3. Adding milk to a cup of hot coffee does not only change its temperature, but also increases its volume. When milk is added, the liquid level in the cup rises, causing the surface area through which heat is lost to increase. This factor favours the milk being added later to reduce the heat loss.

On the other hand, the statement that heat loss by evaporation is proportional to the temperature difference is not true when the temperature of the liquid is well above room temperature. Evaporative heat losses increase more than in proportion to the temperature difference, so these losses are minimised by adding the milk later.

4. Gases and the atmosphere

Photograph of the Earth's horizon, showing the atmosphere as a thin blue line, taken from the International Space Station while orbiting above South America (source: NASA)
http://eoimages.gsfc.nasa.gov/images/imagerecords/50000/50205/ISS027-E-012224_lrg.jpg

We Earthlings live at the bottom of a sea of air, upon which we are entirely dependent for our immediate survival. Because the air is invisible to our eyes, we are rarely aware of its existence or the absolutely critical role it plays in our lives. Humans, our primate ancestors and most living creatures on this planet have evolved to live on the surface of the Earth in a gaseous environment with a pressure of about one atmosphere containing about 20% oxygen. Astronauts travelling into outer space or other planets must wear a space suit or stay within a pressurized capsule to replicate an artificial atmosphere similar to that at the surface of the Earth. Without oxygen, humans and animals could not live more than a few minutes. If exposed to the vacuum of space, we could not survive more than seconds.

The Earth's atmosphere is remarkably thin. Most of the air is contained within ten kilometers of sea-level. This elevation corresponds roughly the top of the "troposphere", within which virtually all weather occurs.

By a height of 100 kilometers, air pressure is one-millionth that at sea level. This can be considered to be the beginning of outer space. Yet even this height is only about 1.5% the radius of the Earth. Seen from outer space, or from an orbiting spacecraft, the atmosphere appears as a thin film covering the surface of the planet. If the Earth were the size of a basketball, the troposphere would be a film less than half a millimeter thick! It would barely cover the dimples on the basketball.

Have a look at the photo taken by astronauts on the International Space Station. I recall reading an account written by a NASA astronaut orbiting the Earth. Her over-riding impression was how extremely thin and tenuous was the film of air upon which we are all so completely dependent.

When the Earth was first formed by the agglomeration of gas and dust swirling around the sun, about 4.5 billion years ago, its atmosphere would have been completely different from what it is today. The Earth's initial primeval atmosphere was probably composed of carbon dioxide and other gases released by volcanoes. As the planet cooled, gas molecules would not have had sufficient energy to escape the downwards pull of Earth's gravity, and began to collect in a layer of gas above the surface. Over millions of years, volcanoes and bombardment by meteorites contributed new constituents to the atmosphere. Geological processes formed new continents and mountain ranges, exposing silicate rocks to rain and weathering and causing carbon dioxide to be absorbed from the air (and bound as calcium carbonate within limestone deposits). With the appearance of photosynthetic bacteria, algae and plants, able to absorb carbon dioxide and release oxygen, the composition of the atmosphere changed dramatically and began to resemble what it is today.

The Earth's atmosphere consists primarily of nitrogen (78%) and oxygen (21%), and about 1% argon and traces of other gases (most notably carbon dioxide, 0.04%). These proportions relate to the composition of *dry* air. Depending upon the temperature and humidity, air contains 1% water vapour.

For air at one atmosphere pressure, molecules of nitrogen, oxygen and other gases comprise about one-thousandth of the volume of the gas. Consequently, a liter of air at atmospheric pressure has roughly one-thousandth as much mass as when it is liquefied at cryogenic temperatures (when the molecules are much more closely packed together).

The Earth's atmosphere is unique compared to other planets in our solar system. The atmosphere of Mars is only about one-hundredth as dense as the Earth's atmosphere, and consists almost entirely of carbon dioxide. Earth's other nearest neighboring planet, Venus, has a crushing hellish atmosphere about ninety times the pressure at the surface of the earth. The temperature at the surface of Venus is 460°C, and the atmosphere consists primarily of carbon dioxide with clouds composed of sulfuric acid mist. Of various space probes sent to the surface of Venus, none survived more than an hour or two.

Most of the population of the Earth live in coastal cities, towns and rural communities, located just above sea level, where air pressure is defined as "one atmosphere". At sea level, air has a density about 800 times less than water. Nonetheless, because the atmosphere is an ocean of air kilometers high, a very substantial force per unit area is required to support its weight. The force exerted by the atmosphere at the Earth's surface is simply the total weight of a column of air extending into the upper atmosphere. Atmospheric pressure varies slightly with weather conditions, but near se level, always has a value of around 10^5 Newtons per square meter. This corresponds to a mass of air of 10,000 kilograms (10 tonnes) for each square meter of area. Air pressure is expressed in units of Pascals (where 1 Pascal is a force of one Newton exerted per square meter). Normal atmospheric pressure is just about 100 kilopascals (where one kilopascal = 1,000 Pascals).

Normal atmospheric pressure is probably far greater than most people realise because we don't normally experience it directly. The outwards pressure exerted by air within our lungs is balanced by the pressure of surrounding air pushing inwards on our chest. Otherwise, our lungs would explode, as would happen to any unfortunate astronaut who suddenly found themselves in the vacuum of space with no space suit. The force exerted by the air on the surface of this sheet of paper is about half the weight of a small car. We don't notice this because it is balanced by an equal and opposite force acting on the other side of the paper.

The curious behavior of gases

Anyone who has ever used a bicycle pump to inflate a tyre or a basketball would appreciate that air, like all gases, is compressible. As we force the gas to occupy a smaller volume, its pressure increases. As we expand a gas, its pressure reduces.

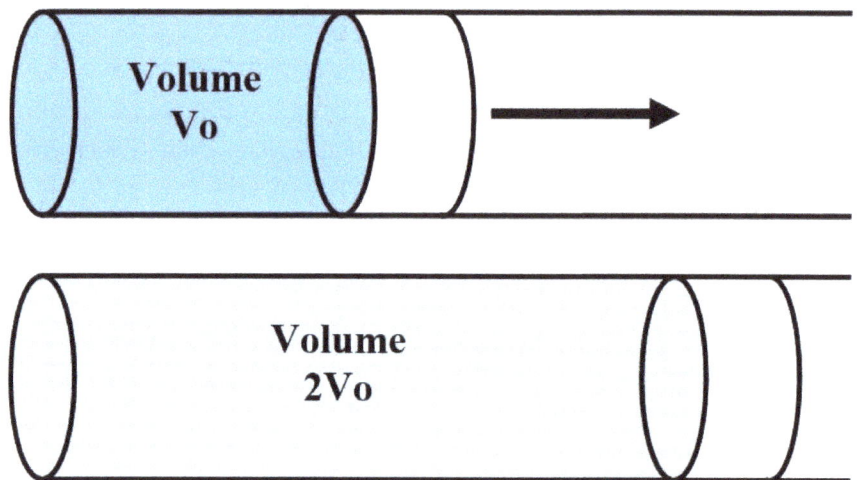

Let's consider a given amount of gas at a constant temperature. We might imagine that the gas is confined within a cylinder fitted with a sliding piston. Let's see what happens to the pressure as we increase its volume by sliding the piston outwards.

Initially, the gas is at pressure **Po** and volume **Vo**. As the volume of the gas increases to twice its initial volume (that is from **Vo** to **2Vo**), the pressure reduces by half. As the volume increases to three times its initial value, the pressure reduces to a third of its initial value.

Thus, for a given amount of gas at constant temperature, the pressure varies inversely with its volume. This relationship was first reported by Robert Boyle (considered one of the founders of modern chemistry), and is called "Boyle's Law".

A short (1-1/2 minute) video demonstrating Boyle's Law for a balloon, marshmellow and shaving cream can be found at: https://www.youtube.com/watch?v=N5xft2flqQU

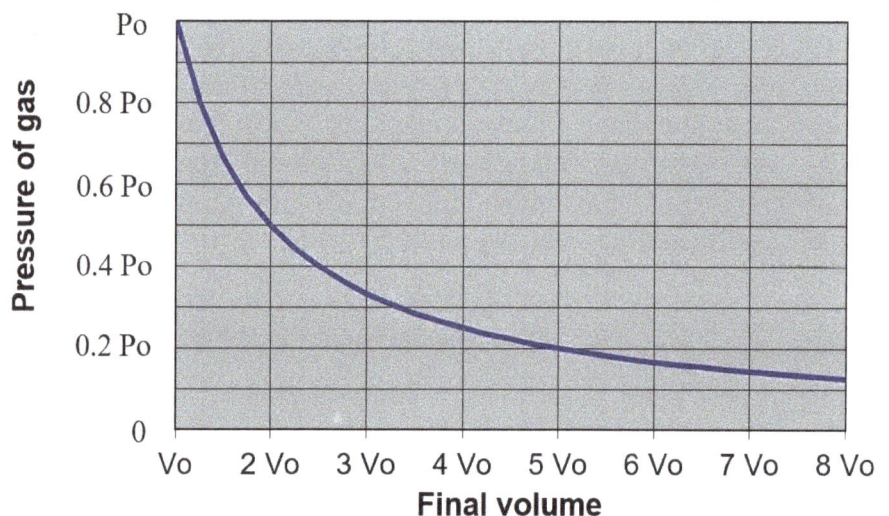

Expansion of a gas at constant temperature

Now let's consider what happens to a given amount of gas as it is heated, while its pressure remains constant. We can do this, for example, by placing a balloon filled with air into a beaker of hot water. As you might perhaps expect, a gas expands when it is heated (at constant pressure). What you might not realise is that the volume of a gas varies in direct proportion with its **absolute temperature** (that is, its temperature relative to absolute zero). Thus, if we were to measure the volume of a gas sample at (say) atmospheric pressure, at several different temperatures, we would find that all of our data points of temperature/volume would fall on a straight line. If we extrapolate this line back to a hypothetical point at which its volume would be zero, this would occur at -273.15°C, absolute zero. I refer to this as a hypothetical point because any gas will liquefy or solidify at temperatures above absolute zero. At very low temperatures, the molecules will be packed as close as they can get, but their volume will not be zero.

Variation of gas volume with temperature
(constant pressure)

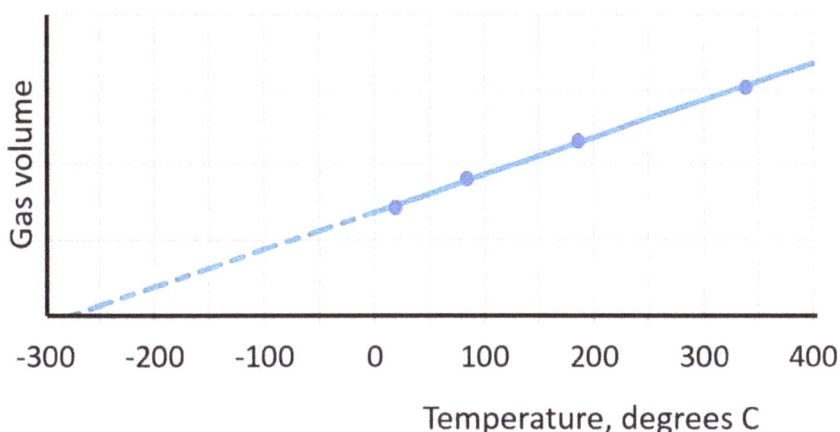

Gas volume vs Temperature, degrees C (-300, -200, -100, 0, 100, 200, 300, 400)

The volume-versus-temperature relationship of a gas is called "Charles' Law", named after Jacques Charles, one of the chemists who discovered it.

Expansion of a gas (air) as it is heated in boiling water is demonstrated in the first short (1-1/2 minute) video below. Even more impressive is the contraction of air that occurs as it cools in liquid nitrogen, shown in the second one-minute video :
1. https://www.youtube.com/watch?v=HQ9Fhd7P_HA
2. https://www.youtube.com/watch?v=Gi5wPnkBEYI

Now that we have some feel for **how** gases behave, it would be good to have an intuitive explanation of **why** gases do what they do. In fact, there is a really simple and elegant theory that explains gas behavior. It allows us to understand why gases obey Boyles Law and Charles Law, and to derive an "Ideal Gas Law". It's called the "kinetic molecular theory of gases".

Why gases behave as they do

Let's leave aside the big view for now, and look at the atmosphere from the perspective of the gas molecules that comprise it.

Imagine that we had incredibly powerful microscopic vision that allowed us to see the gas molecules confined within a container of volume **V**. Bear in mind that, for any gas at atmospheric pressure, there are about 25 billion billion molecules within each cubic centimeter (approximately the volume of your left nostril). Needless to say, this is a huge number of molecules.

It would be fair to say that the number of gas molecules within one cubic centimeter is beyond human comprehension. No human being could count to this number within a single human lifetime, or even many human lifetimes. Try to imagine the number of seconds that have elapsed since the universe came into existence, and then multiply this sixty times!

Rather than trying to deal with such gigantic numbers of molecules, chemists usually talk about "moles" of molecules (just like we might talk about a dozen eggs or a bushel of corn cobs). Each mole contains 602,300 billion billion molecules.

Despite the astronomical number of molecules within a single cubic centimeter, they comprise about one-thousandth of the volume of the container: the remainder being empty space. For example, the size of a nitrogen molecule is about 0.2 nanometers (0.2 billionths of a meter). At room temperature, the average spacing between molecules is about 2 nanometers.

Within a container, air molecules fly around at average speeds of around 500 meters/second (slightly faster than a pistol bullet), constantly colliding with each other. At one atmosphere pressure, nitrogen molecules travel an average distance of 70 nanometers before colliding with another molecule [Note 1]. This distance is called the "mean free path". During a collision, a molecule is likely to be scattered back in the direction from which it came. Thus, after undergoing billions of collisions each second, a molecule is unlikely to migrate more than a few millimeters – even though it travelled hundreds of meters (back-and-forth) within that time.

Molecules near the walls of the container collide with the surface and rebound. Although each molecule has a tiny (seemingly infinitesimal) mass, the container wall must apply considerable force to repulse the onslaught of trillions of molecules striking the wall and rebounding ech second.

For gas confined within a container, trillions of molecules strike the walls of the container each second, and rebound. Consider a small area of the container wall, which is relentlessly bombarded by gas molecules each second. The wall must absorb outwards momentum in bringing these molecules to a momentary stop, and then impart inwards momentum to accelerate the molecules away from the surface. This exerts an outwards force on the container wall.

Molecules striking the walls of the container undergo a change in momentum, exerting an outwards force against the wall of the container. The force per unit area (pressure) depends upon the number of moles of gas per unit volume. The pressure also varies with the average mass of each molecule and the square of its molecular velocity (and hence, its kinetic energy).

When a gas molecule crashes into the container wall, it rebounds. Because of the interchange of energy between the gas and the wall of the container, sometimes gas molecules bounce back at higher velocity and sometimes at lower velocity. On average, however, gas molecules rebound with the same velocity as they impact the wall.

The reversal in velocity – and the change in momentum - of countless gas molecules hitting and rebounding from the wall each second exert an outwards force. The force

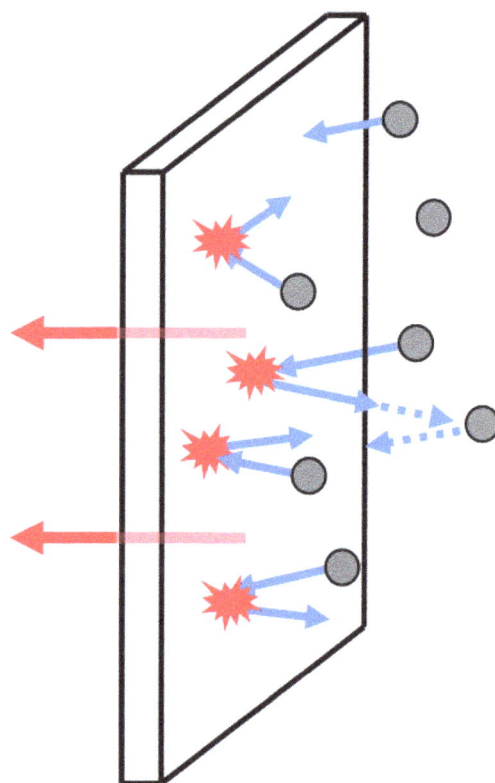

per unit area (pressure) exerted by the gas on the container wall varies directly with:

- The number of collisions per second. This depends on the number of molecules per unit volume. So, if we have **n** moles of gas molecules in volume **V**, the number of collisions varies directly with the ratio **n/V**. The number of molecules hitting the wall also depends on the velocity at which the gas molecules are moving.
- The change in momentum as each gas molecule collides with the container wall. This depends upon the mass of each gas molecule and its velocity.

The number of collisions and the momentum change per collision are each proportional to the average velocity at which the molecules are moving - so the pressure is proportional to *the square of their velocity*. The momentum change is also proportional to the average mass of the molecules. The mass of the molecules and their velocity squared gives the average kinetic energy of the molecules. As it turns out, the average kinetic energy of gas molecules varies directly with the absolute temperature **T** of the gas (that is, its temperature above absolute zero, -273°C). In fact, the average kinetic energy of gas molecules depends *only* on the absolute temperature.

Since the pressure of a gas varies with the ratio **n/V** and with the temperature **T**, we can write the following equation relating the gas pressure to the number of moles of gas **n**, its volume **V** and absolute temperature **T**:

$$\text{Eq (1A)} \qquad P = \left(\frac{n}{V}\right)(\text{constant})(T)$$

Usually, proportionality constants are the least interesting part of equations, but there is something extraordinary and special about the constant in Eq (1A), which is given the symbol **R** and is called the "Universal Gas Constant". Firstly, the value of **R** is indeed *constant for all gases*. Secondly, and even more amazing, R is *universal*: it ends up in all types of equations, including some very important equations that have nothing to do with gases!

We can re-write Eq (5-1A) in the more usual form of Eq (1B), which is termed the "Ideal Gas Law":

$$\text{Eq (1B)} \qquad \mathbf{P\,V = n\,R\,T}$$

> Where **P** is the pressure of the gas, in units of Pascals (Newtons of force exerted per square metre)
> **V** is the volume of the gas, in m³
> **n** is the number of moles of gas
> **R** is the Universal Gas Constant, with a value of 8.3 Joules/mole-°K
> **T** is the absolute temperature in degrees Kelvin (°K = °C + 273)

The "Ideal Gas Law" applies to all gases at "normal" pressures and temperatures that we encounter in the Earth's atmosphere. Under these conditions, gas molecules are sufficiently far apart that they comprise an insignificant fraction of the volume of the container. Furthermore, because the molecules are so far apart, attractive forces between the molecules (responsible for holding the molecules together in a liquid or solid) have negligible effect. The same situation applies reasonably well even at relatively high pressures encountered in compressed air tanks and at the Surface of Venus. The Ideal Gas Law ceases to be applicable only when gas pressures are so high (hundreds of atmospheres) that the density of a gas approaches that of a liquid or solid.

The Ideal Gas Law accounts for a range of gas properties which had been discovered over the centuries. In particular:

- The Ideal Gas Law is consistent with Boyle's Law, stating that, for a given quantity of gas at constant temperature (that is, where **n** and **T** are constant), the pressure varies inversely with its volume.

- The Ideal Gas Law is consistent with Charles' Law, stating that, for a given quantity of gas at constant pressure (where **n** and **P** are constant), the volume varies directly with the temperature.

- The Ideal Gas Law states that, at a given pressure and temperature, the volume of a gas varies directly with number of moles of gas. So, equal volumes of gas (at the same pressure) contain the same number of molecules – regardless of which gases are present. This explains why, when we decompose water (H_2O) into its elements by passing an electric current through it, we get twice the volume of hydrogen gas as of oxygen. Here is a 3-minute video showing the electrolysis of water. You only need to watch the first half: https://www.youtube.com/watch?v=HQ9Fhd7P_HA

Work done by expanding gases

The expansion and compression of gases is fundamental to the technologies of the modern age. It underpins refrigeration and air-conditioning; engines that power our cars, production of electricity in coal, gas-fired and nuclear power stations; and the operation of guns, cannons and rocket engines. Compression of refrigerant gases in air-conditioning units is what enables us to be cool and comfortable on the hottest summer days. Expansion of hot combustion gases, pushing against pistons in engine cylinders, is literally what drives cars, trucks and buses. Expansion of hot superheated steam in steam turbines produces power to generate electricity.

How much work is produced by an expanding gas, or conversely, how much work must be done to compress a gas? We can derive the answer from the Ideal Gas Law.

Let's consider a gas confined within the walls of a cylinder fitted with an air-tight sliding piston. The cylinder has cross-sectional area **A**.

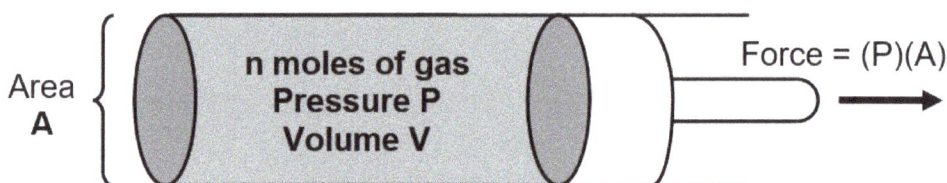

The gas has a pressure **P** pushing the piston outwards. The force on the piston is equal to the gas pressure **P** times the cross-sectional area of the piston **A**.

Let's say that the gas has an initial pressure P_0, an initial volume V_0, and is at temperature **T**.

Imagine that the piston slides a very small distance **x**, causing the volume of the gas to increase by $\Delta V = Ax$. The work done by the expanding gas is equal to the force that it exerts on the piston (**P A**) times the distance **x** that the piston is pushed. As the piston moves, the volume of the gas increases by **Ax**. So, the work done in expanding the gas by a small additional volume ΔV is:

$$
\begin{aligned}
\text{Work done in pushing piston} &= (P\,A)\,x \\
&= P\,(A\,x) \\
&= P\,\Delta V
\end{aligned}
$$

28

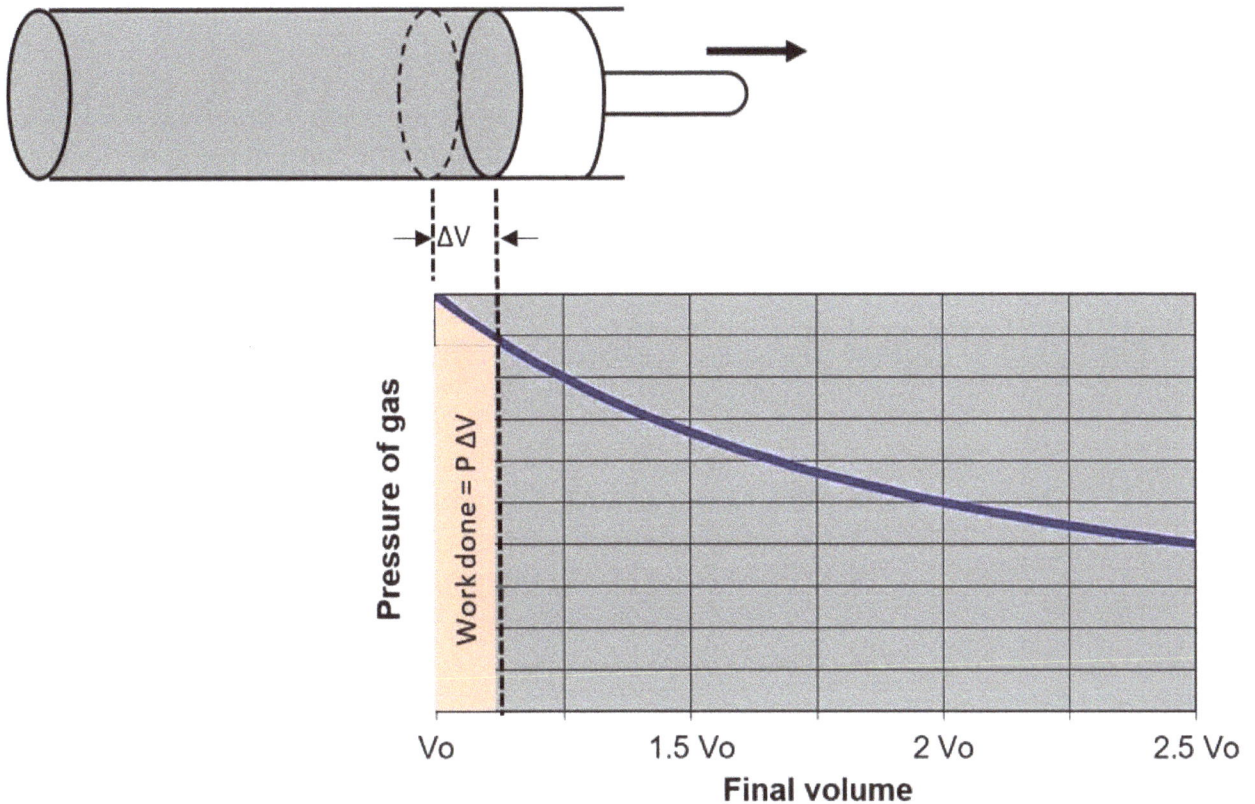

Since distance **x** is much less than the length of the cylinder, the volume of the gas doesn't change very much, so the pressure of the gas hardly changes. Consequently, the work done by the gas is simply the pressure **P** (the vertical height on the pressure-volume curve) times the increase in volume Δ**V** along on the horizontal volume axis. This is simply the area of a thin rectangular segment under the pressure-volume curve.

Now, imagine that the piston slides another very small distance **x**, which again increases the volume of the gas by Δ**V**. Once again, the work done by the gas is **P** Δ**V**. However, since the gas has already been slightly expanded from its original volume, the pressure will be slightly less than its original value. The work additional done by the gas is the area of another thin rectangular segment under the pressure-volume curve.

We can imagine that the gas is expanded in many such small, incremental stages from its initial volume **V**$_I$ to its final **V**$_F$. *The total work done in these many incremental expansions is simply the area under the pressure-volume curve from the initial volume* **V**$_0$ *to the final volume* **V**$_F$*.*

Usually, when a gas is expanded, its temperature reduces. This temperature falls because the work done in expanding the gas is provided by the internal heat energy of the gas. However, in many cases, the reduction in temperature is small or insignificant, and can be neglected. There are two reasons for this. Firstly, if the gas is expanded slowly, it absorbs heat from the surrounding environment. Secondly, if the gas temperature drops by 10 or 20 degrees centigrade, this may seem quite noticeable to us, but the change in temperature is insignificant compared to the *absolute temperature* of the gas. Consequently, there are many applications where we can neglect the change in temperature of a gas when it is compressed or expanded. We consider the temperature to remain constant. This is called an "isothermal expansion" ("isothermal" meaning simply "constant temperature").

If the temperature of the gas remains at a fixed temperature T, then the pressure of an ideal gas is inversely related to its volume by the Ideal Gas Law.

$$P = \frac{n\,R\,T}{V}$$

For a given quantity of gas (n moles), we can plot the pressure-volume graph curve. Using the mathematical technique of "integration", we can determine the area of the graph – and the work done by the gas as it expands from its initial volume V_0 to its final volume V_F.

The result is:

Eq (2)   Work done by expanding gas $= n\,R\,T\,\ln\left[\dfrac{V_F}{V_0}\right]$

Where n is the number of moles of gas
   R is the Universal Gas Constant (8.3 Joules/mole-degree K)
   T is the absolute temperature of the gas
   V_F/V_0 is the expansion ratio, the number of times the volume
       of the gas is increased.

The term $\ln[V_F/V_0]$ asks: "how many times should e be multiplied by itself to give the ratio V_F/V_0. Each time a gas expands 2.7 times in volume, the work done by the gas is nRT.

For one mole of gas, each time that its volume expands 2.7 times, the work done is RT. At room temperature, RT has a value of 2,500 Joules.

Work done by expanding one mole of gas
constant temperature 25 deg C

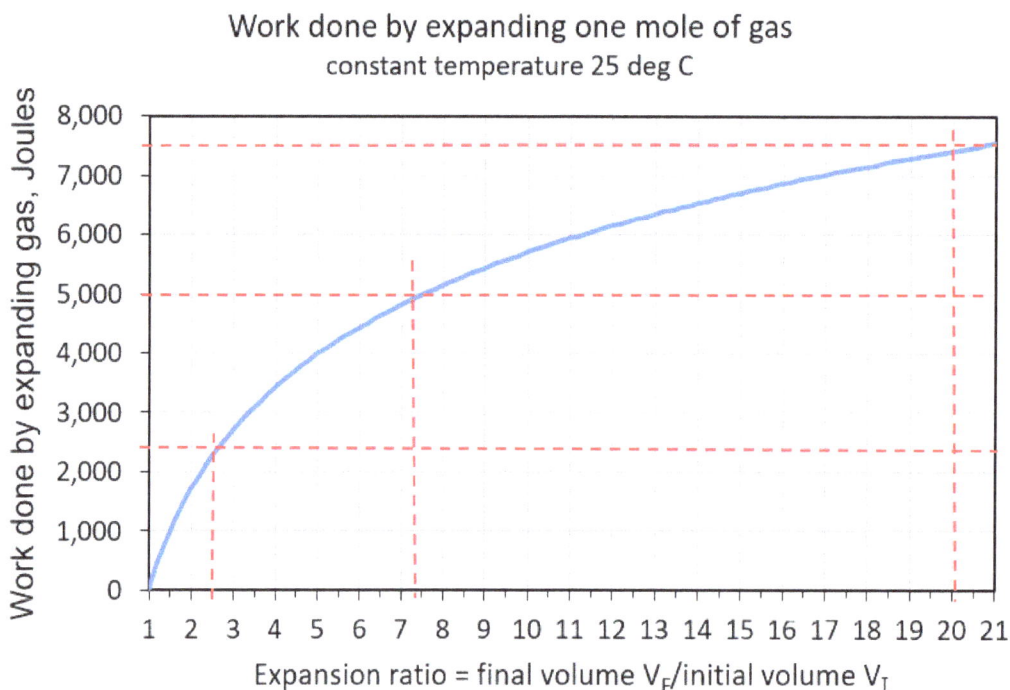

Expansion ratio = final volume V_F/initial volume V_I

Equation (2) shows *the <u>work done</u> by a gas as it <u>expands</u> to a larger volume*. However, it also gives *the <u>work required</u> to <u>compress</u> a gas to a smaller volume*. When a gas is compressed, the ratio of the final volume to the initial volume V_F/V_0 is less than 1.0, so the term $\ln[V_F/V_0]$ is negative (since work is done *on* the gas, rather than *by* the gas). Each time a mole of gas is compressed to about one-third (**1/e**) of its volume, a work input of RT is required.

Notes

1. The average distance that gas molecules travel before they collide with other gas molecules, called the "mean free path", varies enormously, depending upon the pressure.

   At "normal" pressures to which we are accustomed, of about one atmosphere, the mean free path is microscopically small – about the size of a single virus particle. At these pressures, each molecule undergoes billions of collisions with its neighbours each second. Since a molecule is equally likely to scatter in one direction as another, it bounces back-and-forth and doesn't move far from its original position. Gas molecules tend to stay near their neighbouring molecules. If the gas is flowing, molecules are pulled along with their neighbours (like a hapless bystander carried along with the crowd at a football stadium).

   At very low pressures, it is a completely different story. I have worked with high vacuum systems, which routinely maintain gas pressures of about one billionth of an atmosphere. At such pressures, gas molecules travel for distances of several meters before they collide with another molecule. The "mean free path" is generally much longer than the size of the vacuum chamber, so gas molecules rarely collide with each other. Rather, molecules travel in straight lines until they hit a surface within the vacuum chamber. Pumps for high vacuum systems (either "diffusion pumps" or "turbomolecular pumps") must have very large openings, since gas molecules will not "find" the opening to the pump unless they happen to be travelling straight towards it.

   Ultra-high vacuum systems maintain gas pressures of about one trillionth of an atmosphere. At these pressures, gas molecules travel for distances of *kilometers* before they collide with another molecule. In a laboratory-size vacuum chamber, the gas molecules virtually never collide: they travel in straight lines until they hit and bounce off a solid surface.

2. We have considered the situation when a gas expands at constant temperature, called "isothermal expansion". This scenario is accurate and Equation (2) is valid when the expansion occurs slowly, or when the gas volume increases up to three times. Another scenario is for the gas to be expanded with no exchange of heat with its surroundings. This is the case when an expansion occurs rapidly, so there is no time for the gas to absorb heat from its surroundings. In this case, called "adiabatic expansion", the work produced by the expansion is derived entirely form the internal heat of the gas, so the gas cools as it expands. Here is a graph comparing the work produced by one mole of air undergoing isothermal expansion versus the work produced under adiabatic conditions. As you can see, there is not much difference for small expansion ratios, but isothermal expansion produces far more work at high expansion ratios.

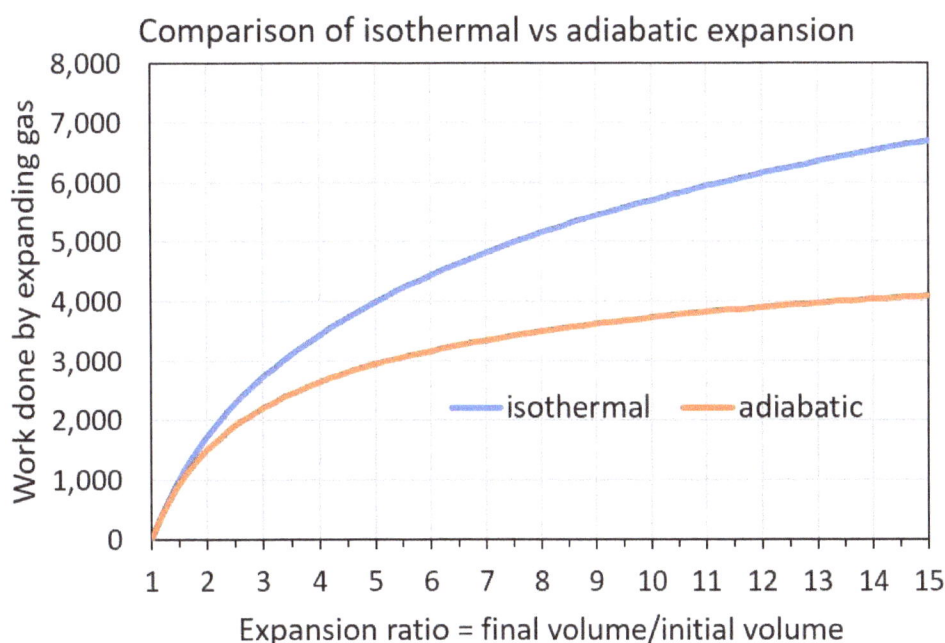

Comparison of isothermal vs adiabatic expansion

5. Variation of pressure with altitude in the atmosphere of Earth and other planets

Because of the gravity of the Earth, gases and liquids are pulled downwards. As a result, the pressure in any fluid increases with depth. Imagine that we are diving down within an ocean of fluid, and let's consider how much the pressure increases with a small increase in depth ΔH.

Imagine a disk-shaped volume of fluid of cross-sectional area **A** and height ΔH. As we have seen before, the mass of this disk-shaped volume is the fluid density ρ multiplied by the volume of the disk **A** ΔH. The weight of this disk is the mass times the acceleration of gravity, or ρ **A** ΔH **g**. This weight is distributed evenly across area **A**, leading to a change in pressure between the top and bottom of the disk.

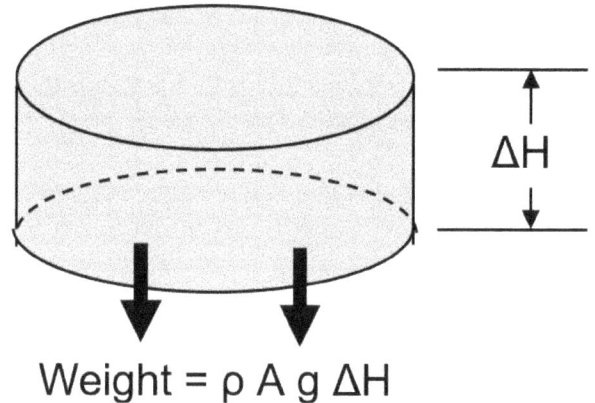

$$\text{Weight} = \rho \, A \, g \, \Delta H$$

Equation (1)   Increase in pressure ΔP in height ΔH = ρ ΔH **g**

This equation is completely general for any fluid, anywhere. For an incompressible fluid, like water, the density is constant and doesn't change with depth. But gases are compressible, and their density varies directly in proportion with the pressure. Let's see why this is:

The density of a fluid is its mass per unit volume. Imagine that we have volume **V** containing **n** moles of gas. According to the Ideal Gas Law, the number of moles of gas per unit volume is:

Equation (2)   Number of moles per unit volume, $\dfrac{n}{V} = \dfrac{P}{RT}$

The mass of each mole of gas is given by its Molecular Mass **M**. For nitrogen, which comprises most of the Earth's atmosphere, its molar mass is 28 grams/mole, or 0.028 kilograms per mole. Air also contains 21% oxygen (and traces of other gases), so it has an average molecular mass of 0.029 kilograms.

The density of a gas is the number of moles per unit volume (given by Equation 2) multiplied by the mass of each mole:

Equation (3)   gas density $\boldsymbol{\rho} = \dfrac{P M}{R T}$

The reduction in gas pressure **ΔP** accompanying an increase in elevation ΔH is given by inserting Equation (3) into Equation (1). The result is:

Equation (4) $\Delta P = P \left[\dfrac{Mg}{RT} \right] \Delta H$

Equation (4) tells us that *the change in pressure with height varies directly with the pressure.*

Imagine that we are slowly rising through the atmosphere in a hot air balloon. Near ground level, where air pressure is greatest, the air pressure falls very quickly with each kilometre increase in elevation. But, as we rise through the atmosphere, the pressure will fall more gradually with each rise in elevation. By the time we reach the altitude at the peak of Mount Everest, air pressure will have fallen to **1/e**, or about one-third, the pressure at sea level.

In fact, for each 8,000-or-so metres that we rise through the atmosphere, the air pressure falls to about one-third of its previous value.

When we rise to twice this elevation, 16,000 metres, air pressure falls to 1/3 X 1/3 = 1/9 the pressure at sea level. At three times this elevation, about 24,000 metres, the air pressure falls to about 1/3 X 1/3 X 1/3 = 1/27th the pressure at sea level.

As we rise higher and higher above the surface of the Earth, atmospheric pressure gets less and less, but technically, never reaches zero. However, by a height of 100 kilometres (which some people consider the beginning outer space), Earth's atmosphere is only one-millionth the pressure at sea level. At an elevation of several hundred kilometres, air pressure becomes essentially zero for all practical purposes.

Since the reduction in pressure with increasing elevation varies directly in proportion with the pressure, the pressure reduces exponentially with height. By applying the mathematical technique of "integration" to Equation (4), we can derive an equation describing how air pressure varies with elevation. This is straightforward for anyone who has studied first-year university calculus. The result is:

$$\text{Equation (5)} \qquad P_H = P_o \, e^{-\left[\frac{Mg}{RT}\right]H}$$

Where P_H is the pressure at elevation **H**
Po is the pressure at ground level (**H** = 0)
R is the Universal Gas Constant (8.3 Joules/mole-deg K)
T is the temperature of the atmosphere
M is the average molecular mass of atmospheric gases
g is the acceleration of gravity (9.8 metres/sec^2 on Earth

Equation (5) tells us that the air pressure reduces by a factor of **e** (roughly, one-third) for each increase in elevation of **RT/Mg**, which is the "effective height" of the atmosphere (roughly, 8,000 metres). This is the height that **would** hold all the air in the atmosphere **if** its pressure were the same as at sea level.

In deriving Equation (5), we assumed that the temperature of the atmosphere doesn't vary with elevation – but this isn't true ! As we rise from the surface of the Earth, air temperature falls by about 6.5 degrees C for each kilometre rise in elevation. However, from the point of view of "the big picture", this doesn't matter very much. Remember that temperature **T** is the **absolute temperature** of the air (relative to absolute zero). To we humans living in Brisbane, a temperature change of 10 or 20°C makes a big difference to our comfort (the difference between summer and winter), but it is not a significant change in absolute temperature. Furthermore, although the temperature of the air changes with elevation in the lower atmosphere (through the troposphere, up to about 10 kilometres elevation), the temperature for the next 20 kilometres of elevation hardly changes at all (and above that, rises very slowly).

Let's compare the results of Equation (5) with the actual variation of pressure with altitude (as given for typical conditions by the US Standard Atmosphere).

Variation of air pressure with elevation

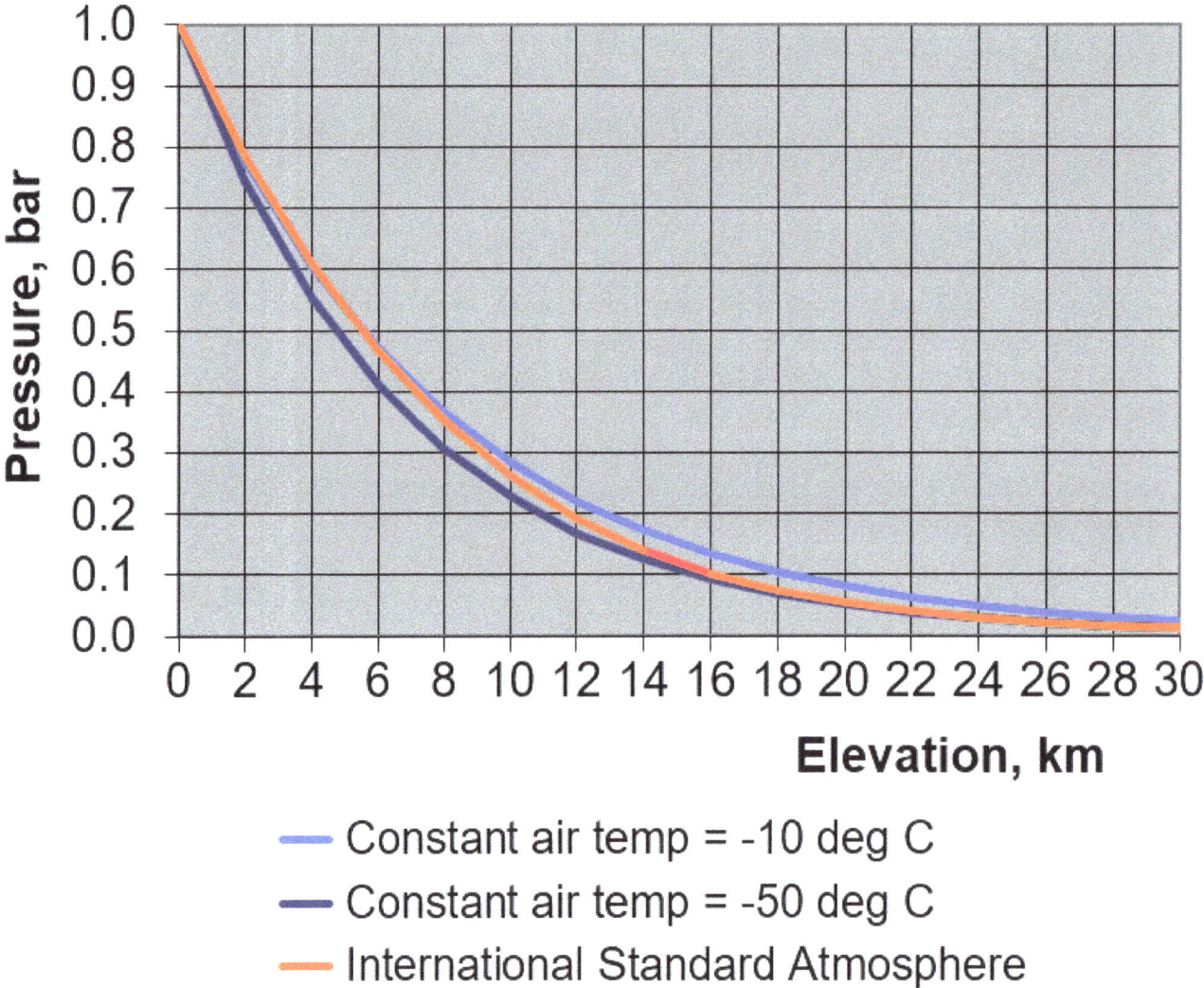

If we assume an average air temperature of -10°C, the (light blue) exponential curve given by Equation (5) is virtually identical to the actual variation in pressure through the troposphere, up to about 10 kilometres elevation. Even at higher elevations, through the stratosphere, the exponential curve of Equation (5) looks very similar to the actual variation of pressure with altitude.

If we assume an average air temperature of -50°C, the (dark blue) exponential curve of Equation (5) is virtually identical to the actual variation in pressure at elevations above 10 kilometres. This should not be surprising, as the air temperature remains nearly constant at -50°C for elevations of 10-30 kilometres.

Another perspective on variation of pressure with altitude

The variation of pressure with altitude, given by Equation (5), was derived by considering how the weight of a column of air varies with height. It is instructive, I think, to derive Equation (5) from a different perspective.

Let's consider the perspective of a bubble of air that is drifting through the atmosphere. For those of us who have made bubbles, we know that bubbles freely drift through the air, rising or falling with the air currents. Because of the weight of the film of soapy water that encloses air inside, bubbles tend to be slightly heavier than the surrounding air (and thus, to gradually sink downwards). However, if we were to make a large bubble enclosed within a very thin film, it would freely drift upwards or downwards with gentle air currents. The liquid film surrounding the air helps us visualise a volume of air drifting within the atmosphere, but is unnecessary for the following arguments.

Although air has low density (about a thousand times less than water), it still has weight. So, for a bubble of air to rise through the atmosphere, an upwards force must be exerted to lift the bubble against its own weight **W**. For the bubble to rise to height **H**, work must be expended for the lifting force **W** to act over distance **H**. So, where does this work come from?

This question can be approached from a number of different viewpoints. Let us now focus on the "viewpoint" of air in the bubble. Of course, a bubble of air doesn't have a viewpoint, but let's pretend that it does. The gas bubble "knows" that work, and energy, are required to lift it against its own weight, but this requires a source of energy. What source of energy is available to our bubble? The answer is: the expansion of the gas to lower pressure as it rises through the atmosphere.

Let's say that the bubble contains **n** moles of air. Each mole of gas has a mass given by its average molecular mass **M** (for air, 0.0285 kilograms per mole). The weight of air in the bubble is its mass **nM** multiplied by the acceleration of gravity **g** (9.8 metres/sec^2 near the surface of the Earth). So, an upwards force **nMg** is needed to push against the weight of the bubble. To raise the bubble to height **H**, this force must be exerted over distance **H**. Consequently, in rising to elevation **H**, the bubble gains potential energy **nMgH**.

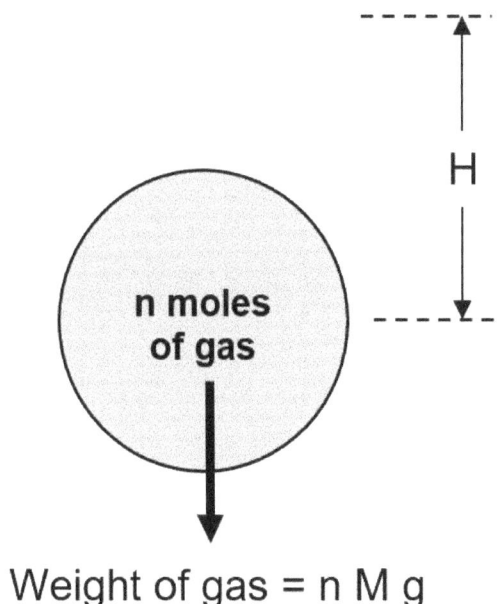

Weight of gas = n M g

We know that the pressure of the air reduces at higher elevation. In fact, *the air pressure **must** reduce so that expansion of the air can provide the work to lift the air within the atmosphere*.

We have previously determined an equation for the work done by a gas as it expands from an initial volume V_o to a final volume V_F. This equation can also be written in terms of the work done by a gas as it expands from initial pressure P_o to final pressure P_F. Since, from Boyle's Law, the pressure varies inversely with volume (for a given amount of gas at constant temperature), the work done by an expanding gas is:

$$\text{Work done by expanding gas} = n\,R\,T\,\ln\left[\frac{P_o}{P_F}\right]$$

35

In rising through the atmosphere, the gas bubble expands as the surrounding air pressure reduces from P_o at ground level to P_H at height H. Let's **equate the work done by the expanding air to its gain in potential energy as it rises** to elevation H. We get:

$$n R T \ln\left[\frac{P_o}{P_H}\right] = n M g H$$

Work done as gas expands and rises   Gain in Potential energy

We cancel out **n**, which appears on both sides of the equation, and re-arrange the equation to find the pressure P_H at elevation H.

We get:

$$\ln\left[\frac{P_o}{P_H}\right] = \frac{M g H}{R T}$$

The term **ln [P_o/P_H]** "asks" us: how many times must **e** be multiplied by itself to give P_o/P_H? In other words, we can rewrite this equation in the form:

$$P_H = P_o\, e^{-\left[\frac{M g}{R T}\right]H}$$

This is exactly the same equation derived before as Equation (5).

This provides an alternative approach to understand the exponential reduction of pressure with altitude. For an atmosphere at constant temperature, **the exponential reduction of air pressure with elevation ensures that, as the gas rises, the work done by the expanding gas supplies the energy needed to lift the gas.**

Conversely, if a bubble of air drifts to lower elevation, the weight of the bubble (and loss of potential energy as it falls) provides the work to compress the gas to higher pressure.

Variation of pressure with altitude on other planets

Equation (5) allow us to determine the variation of pressure with altitude on any other planet or moon (assuming that we know the composition of the atmosphere, and its surface temperature and pressure).

Here is a table that I compiled, containing information about the atmospheres of three rocky planets (the Earth, Mars, Venus) and Saturn's largest moon Titan. Titan is the only moon in our solar system with a dense atmosphere and liquid-filled lakes (liquid methane) on its surface.

	Earth	Mars	Venus	Titan (moon of Saturn)
Gases in atmosphere	N_2, O_2	95% CO_2	96% CO_2	97% N_2 2% methane
Atmospheric pressure at surface	1.00 atm	0.01 atm	91 atm	1.44 atm
Average molecular mass **M** kg/mole	0.0285	0.044	0.044	0.028
Acceleration of gravity at surface **g**, metres/sec²	9.8	3.71	8.87	1.35
Average temperature at planet's surface	15°C	-60°C	460°C	-180°C

The atmospheric composition and gravity of these planets and moon was measured by spacecraft that landed on, orbited or flew by these bodies.

Titan is the only body in the solar system whose atmospheric pressure at the surface is comparable to Earth. An astronaut walking on Titan would **not** need a pressurised space suit, as would be essential for survival on our moon or Mars. Of course, an astronaut on Titan **would** require a source of oxygen to breathe, and insulated clothing and heating system to avoid freezing to death. A spacesuit designed for Titan might well look similar to those worn by astronauts on the moon, but it would serve an entirely different purpose.

Using Equation (5) and the data in the table above, I have plotted graphs showing how atmospheric pressure varies with elevation on Titan and on Earth.

What is striking is that atmospheric pressure reduces more more gradually with altitude on Titan than on Earth. This is primarily due to the much weaker gravity on Titan, where the acceleration of gravity **g** is about one-eighth as much as on Earth. This effect is offset by the much lower temperature at the surface of Titan, which is one-third the absolute temperature on Earth. The weaker gravity on Titan over-rides the effect of its lower temperature, so atmospheric pressure on Titan falls off more slowly with elevation than on Earth. At the surface of Titan, atmospheric pressure is 40% more than at Earth's surface, but at 30 kilometres elevation, atmospheric pressure on Titan is ten times as much as at the same elevation on Earth.

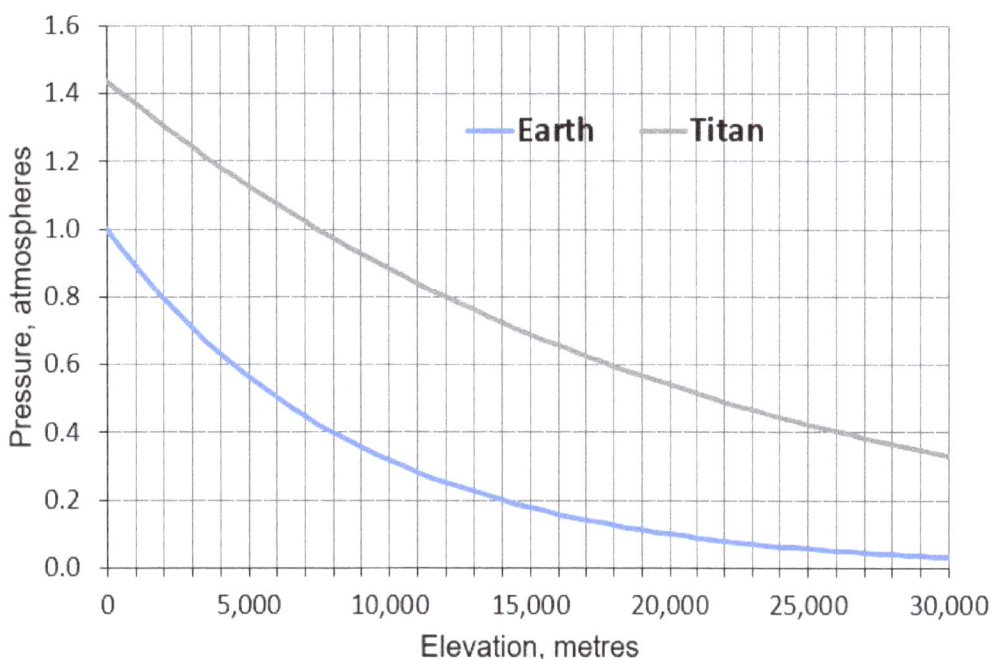

Pressure variation with elevation
Earth versus Titan

Atmospheric pressures on Mars and Venus are vastly different from Earth or Titan. In the case of Mars, atmospheric pressure at the surface is one-hundred times less than on Earth. In the case of Venus, atmospheric pressure is nearly a hundred times more. So, to compare how atmospheric pressure varies with elevation, I have plotted the **ratio** of the pressure at elevation **H**, compared to the pressure at the surface, for Earth, Mars, Venus and Titan.

On Mars, the atmosphere is very, very thin near the surface, but the pressure reduces more gradually with elevation than on Earth due to Mars' weaker gravity.

Atmospheric pressure on Venus reduces even more gradually with elevation, but this is not due to the weaker gravity on Venus (which is only slightly less than on Earth), but to the much higher surface temperature of Venus.

Pressure variation with elevation
for various planets and moons

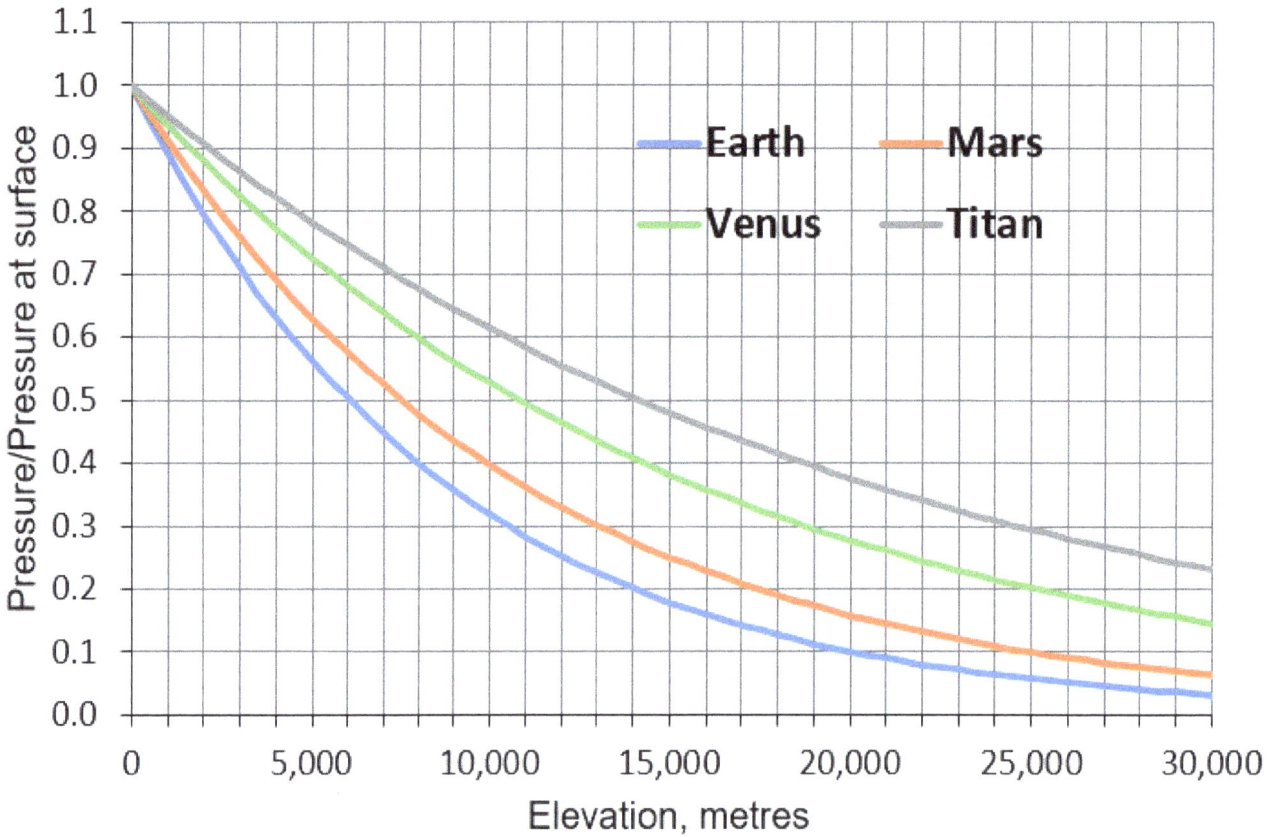

Venus has very high atmospheric pressure (90 times greater than on Earth), and very high temperatures, at its surface. On Venus, to get the same atmospheric pressure as found at the surface of the Earth, we would need to rise to an elevation of about 50 kilometres. At this elevation, we would also find that the temperature is similar to that experienced at the surface of the Earth. So, an elevation of 50 kilometres is likely to be the only place on Venus that would be remotely suitable for humans to survive. Some scientists believe that the best way to explore Venus would be from a balloon drifting through the Venusian atmosphere at 50 kilometres altitude.

6. What the atmosphere tells us about what happens on Earth

As we rise from ground level to higher elevation, and as the pressure reduces exponentially, there are fewer and fewer gas molecules per unit volume.

The exponential reduction in the number of molecules with elevation reflects a fundamental feature of all molecules - *the number of molecules reduces exponentially at higher energy*. In the particular situation in the atmosphere, molecules at high elevation have greater potential energy. More work is required to lift a molecule to greater height, and more work is required to lift a heavier molecule.

Imagine one mole of gas molecules drifting through the atmosphere. The mass of gas is simply the molecular mass **M**, and its weight is **Mg.** At height **H** above the ground, its potential energy is **MgH**. This is the work that must be done to lift one mole of air from ground level to height **H**. Consequently, we can rewrite the equation relating variation of pressure with altitude in terms of the number of molecules N_E with gravitational potential energy **E**, compared to the number of molecules at ground level (N_o).

Equation (1) $$N_E = N_o e^{-E/RT}$$

Not only does atmospheric pressure decline with height, but *the number of molecules declines exponentially with their gravitational potential energy*.

Let's look at this from the perspective of a single molecule, which we might call Fred. Upon being introduced into the atmosphere, Fred undergoes numerous collisions with other molecules, gaining or losing energy in each collision. In some collisions, Fred gains kinetic energy from an impacting molecule, allowing him to rise to higher elevation. In other cases, Fred loses energy to the other molecules.

One reason why gas molecules can either gain or lose energy during each collision is due to the difference in mass of the molecules. Imagine that Fred is a molecule of argon (which comprises 1% of the molecules in air). Fred will frequently collide with nitrogen molecules, which are less massive than he is. Consequently, even if a colliding nitrogen molecule has the same kinetic energy as Fred, it will be moving faster. A head-on collision between Fred and a nitrogen molecule is similar to a collision between a cricket bat and a ball: the more massive object (the bat) transfers some of its kinetic energy to the lighter object (the ball). In such a collision, the more massive molecule loses energy. On the other hand, in a tail-end collision, a heavy argon atom is struck by a faster nitrogen atom moving in the same direction. In this type of collision, the more massive molecule gains energy.

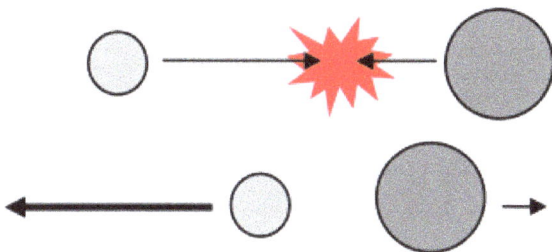

A head-on collision transfers energy from the more massive to the lighter molecule.

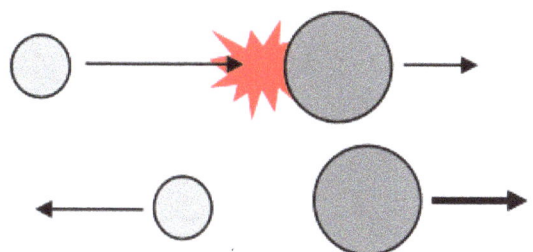

A tail-end collision transfers energy from the lighter to the more massive molecule.

The outcome of a collision, whether a molecule gains or loses energy is random. Whether a molecule "wins" or "loses" is a game of chance, like flipping a coin or placing a bet at the casino.

After numerous collisions, argon atoms will vary widely in energy, and so will nitrogen molecules. There is no maximum limit on how much energy a molecule can gain, although it becomes increasingly unlikely to find a molecule as we look at higher and higher energy. To acquire very high energy, a molecule would need to "win" during one collision after another, for many collisions. Anyone who has patronized a casino knows that the chance of having many wins in a row is unlikely, and the probability of leaving the casino with a huge win declines with the size of the jackpot.

The result of many collisions is that energy is not equally distributed among molecules. And, since molecules are colliding all the time, the energy of any particular molecule changes over time. At any one instant, Fred might have more energy than average, but at other times, Fred will have less. Occasionally, Fred may experience a "winning streak", gaining energy each time he collides with other molecules. At that time, Fred will literally be "riding high", with enough energy to cruise high up in the atmosphere. However, the probability of such a winning streak is low and the gains are temporary.

As Fred wanders through the atmosphere, exchanging energy with other molecules, he will occasionally find himself at 30 kilometres elevation, but it is much more likely that he will find himself near ground level. Over an extended period of time, Fred will find himself at every possible elevation, and every possible potential energy **E**. Since Fred is no different from countless trillions of molecules in the atmosphere, it should be apparent that **Equation (1) reflects the probability of finding any one molecule** at elevation **H** and **with gravitational potential energy E**. In Equation (1), N_E represents the relative probability of a molecule having potential energy **E**, compared to the probability N_o of having no potential energy (that is, being at ground level).

The gravitational potential energy of a molecule varies from zero (at the Earth's surface) to very high values (in the upper atmosphere). If we survey the total population of gas molecules throughout the atmosphere, we find that the **average** gravitational energy per mole is **RT**, which can be derived from Eq (1).

If we plot the relative number of molecules that have energy **E**, as given by Equation (1), we get the graph shown here.

The region just above the Earth's surface has very little gravitational potential energy, and is the most densely populated (has the highest pressure).

Energy distribution of molecules

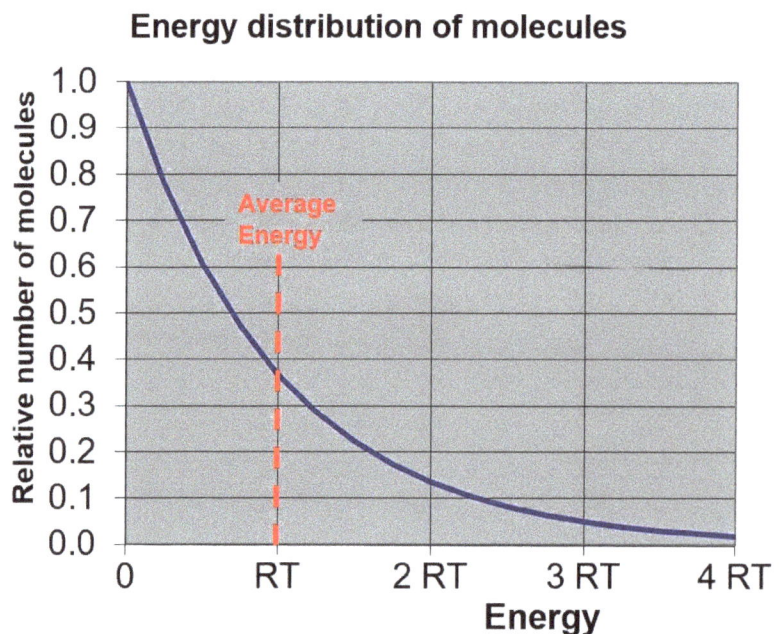

40

As we rise to higher elevation, and higher potential energy, the number of molecules decreases rapidly. About a third of the molecules (1/e = 37%) have more potential energy than the average **RT** (as given by the area under the graph for all energies above **RT**, compared to the total area under the graph). A small fraction of molecules (0.37 X 0.37 X 0.37), or about 5%) have at least three times the average energy, and a tiny fraction (about 0.004%) have at least ten times the average energy.

Equation (1) is not limited to the potential energy distribution of gas molecules in the atmosphere. It turns out that this exponential energy distribution is completely general. Equation (1) is simply a special case of the "Boltzmann Equation" [Note 1], which is fundamental to all chemical processes. The Boltzmann Equation is not limited to gas molecules, and is not limited to potential energy. It applies to the kinetic energy of molecules, their rotational energy, vibrational energy, hydraulic energy, and any other kind of energy. The Boltzmann Equation tells us how energy is distributed amongst a large number of molecules in virtually **any** system in equilibrium at temperature **T**.

To understand why this should be, let's imagine a "thought experiment" in which we create an atmosphere of gas on an imaginary planet, which me might call Glycon (it rolls nicely off the tongue, and sounds vaguely exotic). Since this is a thought experiment, we can choose whatever conditions we like. To keep things very simple, let's assume that there are only 10 molecules in the atmosphere. Each molecule can find itself at any elevation. However, because the molecules are in thermal equilibrium at some particular temperature, **the total energy shared by all the molecules is fixed** at some value, say 100 units.

Imagine that we shake up the 10 molecules, and throw them randomly into the atmosphere. Each molecule will find itself at whatever elevation (and energy) that it lands in. Let's consider how the molecules might distribute themselves through the atmosphere. But remember, that the total energy of the molecules is 100 units, so only those distributions with a total energy of 100 units are allowed.

Since we have 10 molecules, and their total energy is 100 units, the average energy of the molecules is 100/10 = 10 units. Thus, in the first instance, we might imagine that all 10 molecules could fall into the layer possessing 10 units of gravitational energy. But, it is extremely unlikely that this distribution would arise from a random exchange of energy among molecules. Why? **There is only one possible way for this distribution to be achieved**. Imagine that the molecules were assigned numbers. For this distribution to be achieved, Molecule Number 1 would be assigned to the 10 energy unit level, Molecule 2 would be assigned to the same energy level, Molecule 3 would be assigned to the same level, and so on for all 10 molecules. Expecting this situation to arise would be like flipping a coin ten times and expecting to get 10 heads (actually, it is even more improbable than this. It would be like throwing a ten-sided die ten times, and getting the same result each time).

Gravitational
energy

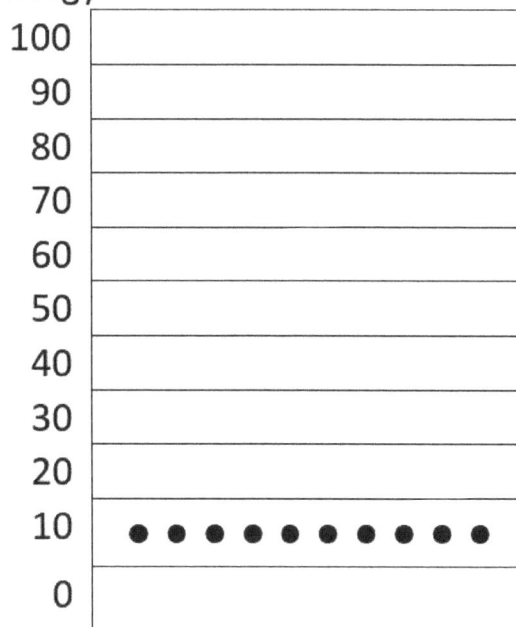

100	
90	
80	
70	
60	
50	
40	
30	
20	
10	● ● ● ● ● ● ● ● ● ●
0	

A more likely situation would be for 9 of the molecules to be in the zero-energy-level ("ground state"), allowing one molecule to be in the 100-unit energy level. This distribution is ten times more likely to occur through a random distribution of energy, since **there are 10 ways for this distribution to occur**. One way is for Molecule No. 1 to be in the 100-unit energy level. A second option - which is equally likely - is for Molecule 2 to be in the 100-unit energy level. A third option is for Molecule 3 to be in the 100-unit energy level. Since there are 10 molecules, there are ten ways to place one molecule in the higher energy level. Consequently, *** this distribution is ten times more likely to occur through a random distribution of energy***.

Gravitational energy

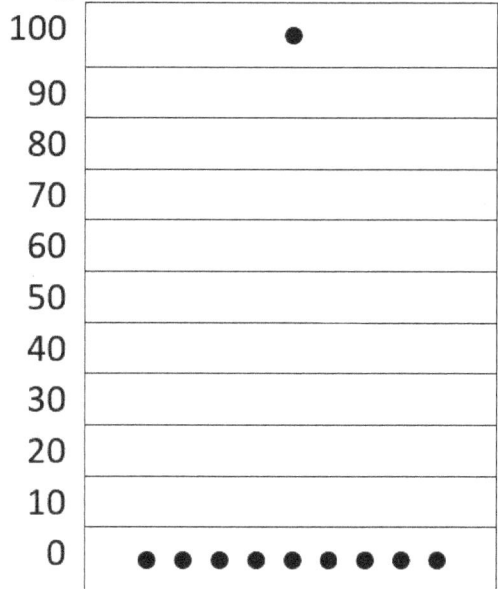

But some other distributions are much more likely than this. We could have 8 molecules in the zero-energy state, and two molecules in the 50-energy-unit level. There are 10 possibilities in choosing the first molecule to go in the 50 energy unit level, and 9 possibilities remaining for the second molecule in this level. It would appear that there are (10)(9) = 90 possible ways to achieve this distribution, but we would be double-counting. If, for example, Molecule No. 3 is the first molecule to land in the 50 energy unit level, and then molecule 7 is the second molecule in this level, this is the same result as occurs when Molecule 3 lands first and then Molecule 7 lands next. The order that the two molecules land is irrelevant. So, in fact, there are (10)(9)/2 = 45 possible ways for this distribution to arise by a random redistribution of energy.

Gravitational energy

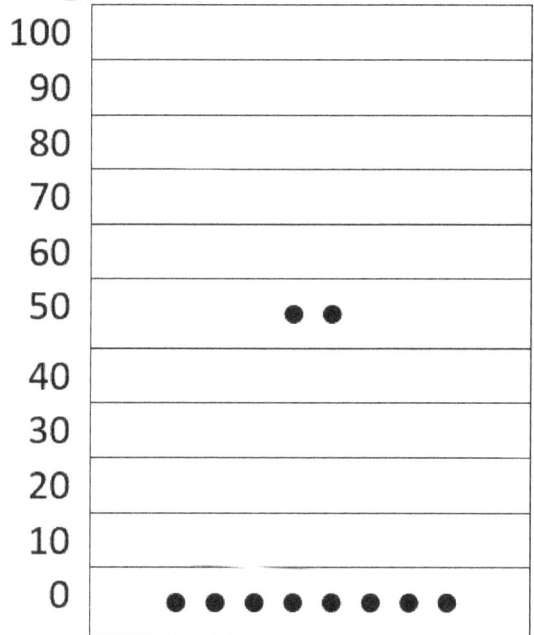

Rather than put two molecules in the 50 energy unit level, we could distribute these two molecules into two energy levels, whose total adds up to 100 units. For example, we could put one molecule in a 60 unit level, and the other in a 40 unit level. This would double the number of possible ways to achieve this distribution to (10)(9) = 90.

By now, you might be starting to recognise a pattern. To achieve the most probable situation, *the molecules should be spread among the greatest number of energy levels, but* to remain within the fixed energy allotment, the molecules *must be concentrated among levels with lower energy*.

Gravitational energy

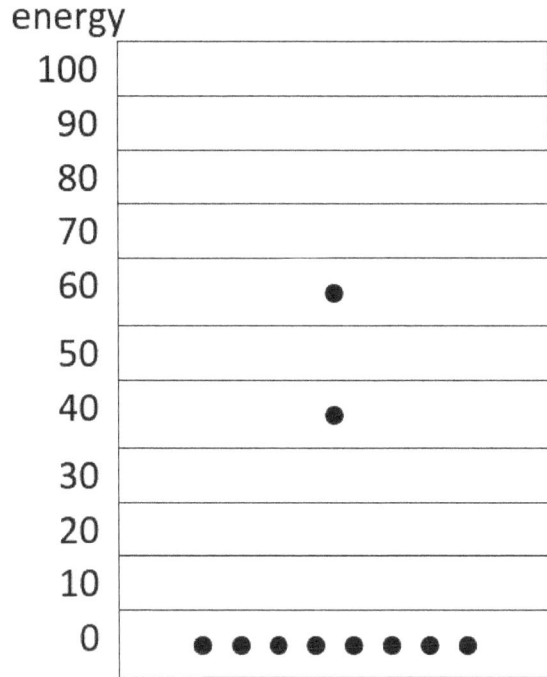

Gravitational energy

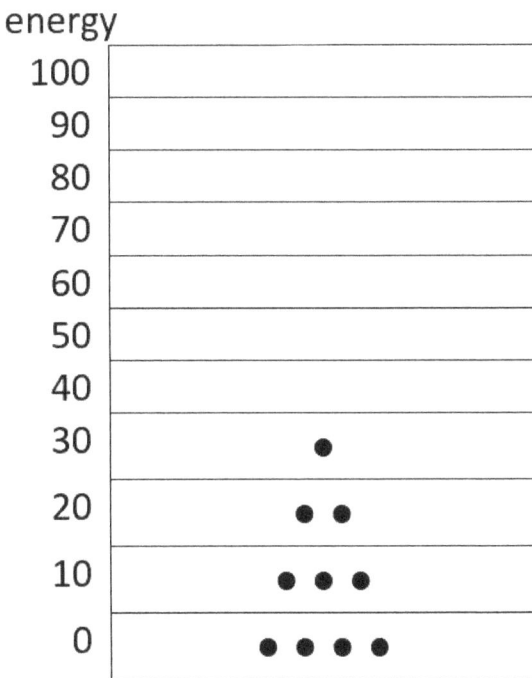

So, we might consider a distribution like the one shown here. I leave as an exercise for the reader to show that **there are** (10)(9)(8)(7)(6)(5) / (3)(2)(2) = **12,600 possible ways to achieve this distribution**. This distribution is far more likely to be achieved by random exchange of energy among the molecules than the previous distributions we looked at.[Note 2]

We have considered a very simplistic situation with only ten molecules. If we repeated this procedure with a large number of molecules, we would find that the most probable situation occurs when the number of molecules reduces exponentially with increasing energy – exactly as given by the Boltzmann Equation.

In fact, the exponential distribution given by the Boltzmann Equation can be derived using exactly the same reasoning used here, and applying several clever mathematical techniques which are applicable when the number of molecules is very large.[Note 3]

The Boltzmann equation gives the distribution of energy that best satisfies the two requirements that molecules are "looking for". Because the exchange of energy among molecules is random, it is extremely unlikely to find molecules jammed together into one small space or one energy level. Molecules distribute themselves with the widest possible variation of energy - but within their fixed total energy. To achieve the widest distribution of energies, within a fixed total energy allocation, molecules tend to congregate among the lower energy levels. They are seeking the widest possible energy distribution allowed by the limited energy available. You could say that the molecules are seeking the greatest randomness or "freedom", or as chemists say, the greatest "entropy".

But this demand for freedom is constrained by the fixed amount of energy available to be shared among the molecules. If very little energy is available (say, at temperatures near absolute zero), the molecules have little choice but to cluster together within lower energy levels.

We can imagine two scenarios:

- **At low temperature** - with little energy available – most molecules are confined near the lowest-possible "ground state" energy level. The limited availability of energy stymies the molecule's ability to occupy higher energy levels.

- **At high temperature** - with ample energy available - molecules can spread out among the available energy states. Each molecule has great freedom to do its own thing, and go wherever chance takes it. We find molecules in different places, moving with different speeds in different directions. Anyone expecting the molecules to all be "well-behaved" and doing the same thing would say that the situation is very disordered. Chemists would say that that the molecules are in a state of high entropy.

The Boltzmann distribution represents the optimal way for molecules to distribute energy amongst themselves in the most random, disordered way that is allowed by their total amount of energy.

Readers may note that the Boltzmann Equation simply gives *the most probable distribution*. You might argue that other energy distributions with the same total energy are possible. And, yes, technically speaking, this is absolutely correct. If we had 10 molecules, it would be possible for all 10 molecules to all have the same energy, as given by our least probable energy distribution. But this is very unlikely. If we flipped a coin 10 times, we would be very surprised to get 10 heads, and no tails. The chance of this occurring is less than one in a thousand. However, even the tiniest sample of a gas or liquid contains *trillions* of molecules. If someone told you that they flipped a coin even one million times (which would take about two months if they didn't sleep, eat or do anything else) and never got a tail, you could be quite confident that they were lying, deluded or using a defective coin. For trillions of molecules, the chance that their energy distribution will differ even slightly from that given by the Boltzmann Equation is infinitesimal. We can be very confident that energy distribution of molecules will follow the exponential fall-off with energy given by the Boltzmann Equation.

You may be wondering why the energy distribution of molecules is so important. It turns out that there are many situations in which molecules face a critical energy barrier that prevents them from reacting or escaping. This applies, for example, to molecules escaping from a liquid into the vapour state during evaporation. In order to escape the liquid, molecules must have sufficient energy to overcome attractive forces to neighbouring molecules. This amount of energy, called the "heat of vaporization" E_{vap}, is absorbed when each mole of liquid evaporates. Any molecule whose energy exceeds this critical amount can pass through the barrier. On the other hand, any molecule whose energy is less than this critical amount is unable to escape, and remains trapped within the liquid.

Molecules that can escape from the liquid are those in the high-energy "tail" of the Boltzmann energy distribution curve. This is why evaporation cools the liquid. Only high-energy molecules escape into the vapour, leaving lower-energy molecules remaining in the liquid.

Energy distribution of molecules

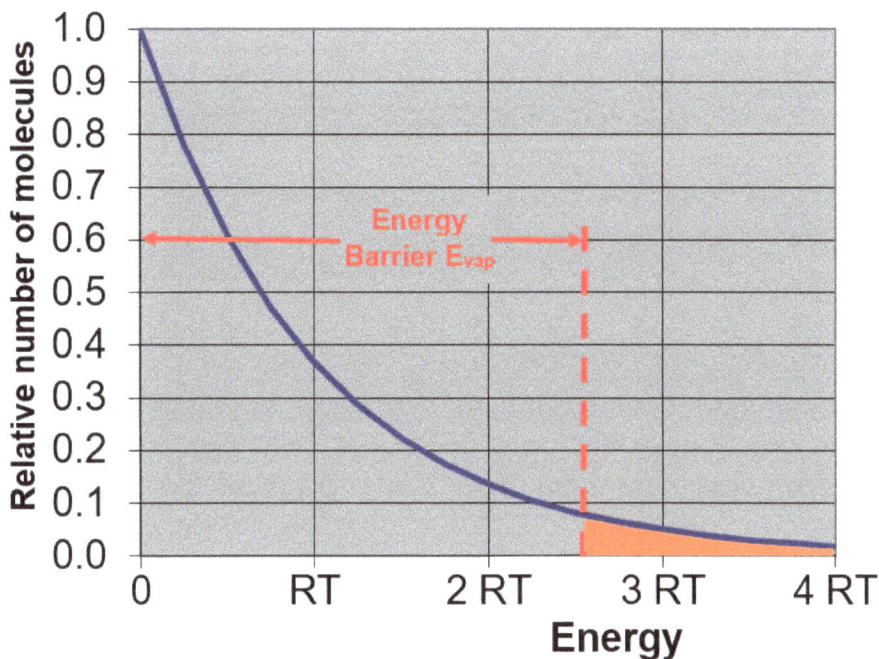

We can determine the fraction of molecules whose energy exceeds the heat of vaporisation E_{vap} by, once again, using the mathematical technique of "integration" to find the area under the energy distribution curve between energy E_{vap} and infinity. When we do this, we get an amazingly simple and elegant result:

Equation (2)   Fraction of molecules exceeding energy $E_{vap} = e^{-E_{vap}/RT}$

Where E_{vap} is the Heat of vaporisation, which is the critical
energy barrier for molecules to escape the liquid.
R is the Universal Gas Constant
T is the absolute temperature

In many cases, the energy barrier E_{vap} confining the molecules is many times greater than their average energy **RT**. Then, only a tiny fraction of molecules (often, one among thousands) have sufficient energy to overcome the energy barrier. In this case, the number of molecules that can react or escape depends critically upon the temperature. Typically, a small increase in temperature (of 10° - 20°C) will cause a reaction to occur twice as quickly. Any chemical reaction or process occurs more quickly at higher temperature.

As it turns out, for nearly all liquids at their normal boiling points, the energy barrier to escape (E_{vap}) is roughly ten times as much as the heat energy of the molecules (RT). This applies equally for liquid butane (boiling at 0°C) or liquid magnesium metal (boiling at 1,090°C). Consequently, even at a liquid's boiling point, only about one molecule in a thousand has sufficient energy to escape into the vapour. However, a relatively small increase in temperature allows a much larger number of molecules to escape the liquid. A 7% increase in absolute temperature (about 20°C at the normal boiling point of butane, or 75°C at the boiling point of magnesium metal) causes about twice as many molecules to escape.[Note 4]

Notes

1. The "Boltzmann Equation" is also referred to as the "Maxwell-Boltzmann Equation, named after James Clerk Maxwell and Ludwig Boltzmann.

2. If we have a total of **N** molecules, distributed with n_1 molecules in the first energy level, n_2 molecules in the second energy level, and so forth, then the number of ways that distribution can be achieved is given by:

$$\text{Number of ways to achieve distribution} = \frac{N!}{n_1!\, n_2!\, n_3!\, \ldots}$$

Where the symbol **N!** ("N factorial") represents $(N)(N-1)(N-2) \ldots (1)$

3. The mathematical derivation of the Boltzmann Equation is rather complicated, and is given in some Physical Chemistry textbooks. The derivation uses the "Stirling Approximation", which approximates the value of ln(N!) where N is a large number. It states that:

$$Ln(N!) = N \ln(N) - N$$

The Stirling approximation becomes increasingly accurate for larger and larger values of N. Since we are concerned with situations involving trillions of molecules, the "Stirling Approximation" is no longer an approximation, but essentially provides an exact value of ln(N!).

4. Since the rate of escape of liquid molecules varies with $e^{-Evap/RT}$, we can derive how the rate of evaporation varies with a small increase in temperature ΔT. The result is:

$$\frac{\text{Rate at temperature } (T + \Delta T)}{\text{Rate at temperature } T} = e^{\left[\frac{E_{vap}}{RT}\right]\left[\frac{\Delta T}{T}\right]}$$

For most liquids, the energy of vaporization E_{vap} is typically about ten times **RT**, so when ($\Delta T/T$) is one-tenth, the rate of escape increases by 2.7 times. The rate of vaporization roughly doubles when the absolute temperature increases by 7% ($\Delta T/T = 0.07$).

7. The world of water and liquids

Evaporation of water, and condensation of water vapour, play a dominant role in shaping the surface of our planet and the lives of plants, animals and human civilisation. Most (about 97.5%) of the estimated 1.3 billion cubic kilometres of water on Earth is contained in the oceans, which has a high salt content. However, humans, animals and nearly all plants grown for food require fresh water. The provision of fresh water (in glaciers, lakes, rivers and underground aquifers) relies on evaporation of water from the sea, transport of water vapour through the atmosphere, and its condensation into rain or snow.

Water is continually cycled between the oceans and fresh water through evaporation and condensation as rain. This plays a major role in the Earth's climate by transferring heat around the planet (from the equatorial to polar regions) and by absorbing heat during the day and releasing heat at night. It also sculpts the surface of the Earth, eroding mountain ranges and forming river valleys, canyons and other features.

Earth is very much a "water world" but, depending upon conditions of temperature and pressure, countless other substances undergo transitions between liquid and vapour. In this respect, water is no different from millions of other chemical compounds. The critical role played by evaporation and condensation of water in creating Earth's climate and surface features could be played by other liquid compounds on other planets.

In 2004, the Cassini-Huygens spacecraft undertook radar mapping of the surface of Saturn's moon Titan and discovered lakes composed of liquid methane, and signs of erosion and river deltas formed by flowing liquid methane (at temperatures of around -180°C). On Titan, methane appears to serve the same role in redistributing heat, creating the climate and sculpting the surface as the water cycle does on Earth.

Ligeia Mare is one of many seas and lakes containing methane and ethane in Titan's north polar region. This image is based on radar mapping of the moon's surface by the Cassini spacecraft in 2006-2007. Source: NASA/PL-Caltech/ ASI/Cornell. (http://photojournal.jpl. nasa.gov/catalog/PIA17031)

In 2014, scientists studying a brown dwarf (an under-sized star, too small to maintain nuclear fusion in its core, with a temperature of about 1,000 degrees Kelvin at its outer surface), found evidence of clouds containing droplets of iron. They believe that "rain" of molten iron falls through the atmosphere, and then evaporates at lower elevations.

Prisoner inside a water droplet

Molecules in a droplet of water, or any liquid, are attracted to surrounding molecules by intermolecular forces. The molecules can slide past one another and move within the liquid - a bit like passengers on a crowded train, with some trying to reach the door at their station. After moving **within** the liquid, a molecule will have different neighbours, but will still have the same **number** of neighbouring molecules. So long as a molecule remains within the liquid, it experiences the same attractive forces to its neighbours.

However, consider the situation of a molecule that tries to escape from the surface of the liquid into the surrounding air or gas. It must overcome strong forces binding it to neighbouring molecules in the liquid.

In the case of water, considerable energy is required for molecules to overcome these attractive forces and escape from the liquid into the surrounding air or gas. About 40,000 Joules of energy are required to vaporise a mole of water. This is called the "molar heat of vaporisation" of water, for which we can use the symbol E_{vap}.

This is a large energy barrier blocking the escape of water molecules to the wider world beyond. Imagine a single water molecule - let's call him "Fred" - inside a liquid droplet peering longingly at the surface and dreaming of a life of freedom in the vast world beyond. Fred's yearning for freedom might seem hopeless. Fred and his fellow water molecules are moving and vibrating, but their average energy (**RT**, or 2,500 Joules per mole at room temperature) is less than one-tenth of the energy required to escape from the liquid. How could Fred hope to get enough energy to escape from the confines of the liquid?

The great escape

But all is not lost! There is a way for water molecules to escape the liquid state.

A drop of water contains about 100 billion billion molecules. These molecules are continually exchanging energy by colliding with one another (and also by emitting and absorbing infrared radiation). Exchange of energy between molecules causes some to have more energy than average while other molecules have less. We have seen that a small fraction of molecules has much greater energy than their colleagues, with the number of such molecules reducing exponentially with energy. At normal room temperature, only a miniscule fraction of water molecules have sufficient energy to escape from the liquid to the gas (vapour).

Only the most energetic water molecules can escape from the liquid into the vapour state, leaving behind molecules that have lost energy during intermolecular collisions. Consequently, the escape of the most energetic molecules from the liquid – which we call "evaporation" – lowers the average energy (and thus, the temperature) of the remaining liquid. Evaporation cools the liquid. Each mole of water that evaporates carries away 40,000 Joules, the Molar Heat of Vaporisation.

The fraction of molecules containing sufficient energy to escape varies exponentially with temperature. Typically, each increase in temperature of about 10°C enables about twice as many molecules to escape into the vapour.

Number of water molecules whose energy exceeds E_{vap} (per million)

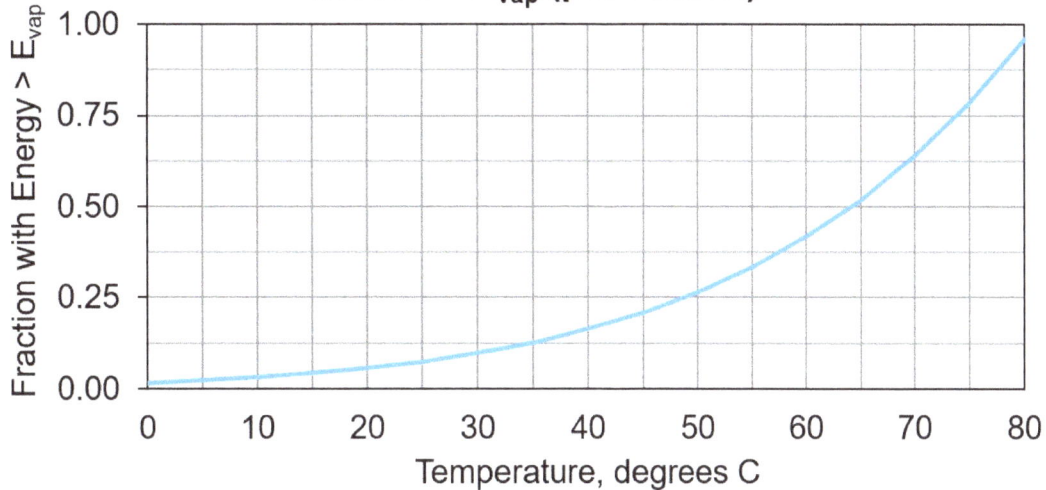

Consider a "thought experiment" in which a sample of liquid water is injected into a sealed container which initially contains no gas or vapour. Initially, before the water is injected, the pressure inside the container is zero.

A tiny fraction of water molecules has sufficient energy to escape from the liquid and wander freely in the space above the liquid. Over time, more and more molecules escape to the gas phase, and the gas pressure increases.

However, as more water molecules occupy the gas phase, some find themselves immediately above the liquid surface and are re-captured.

An equilibrium is soon reached at which water molecules escape from the liquid at the same rate as water molecules in the gas phase return to the liquid. Then, the gas pressure remains steady at the "equilibrium vapour pressure".

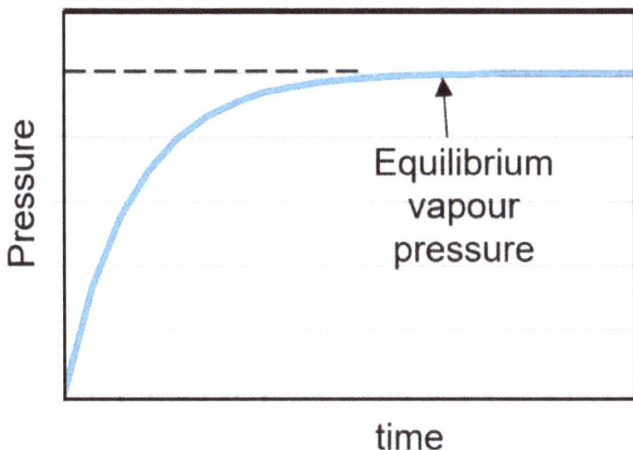

Pressure gauge

A graph of the gas pressure versus time would look like the one shown here. If no air molecules are present (to get in the way of escaping water molecules), the rise of pressure and then its plateau, occurs extremely quickly.

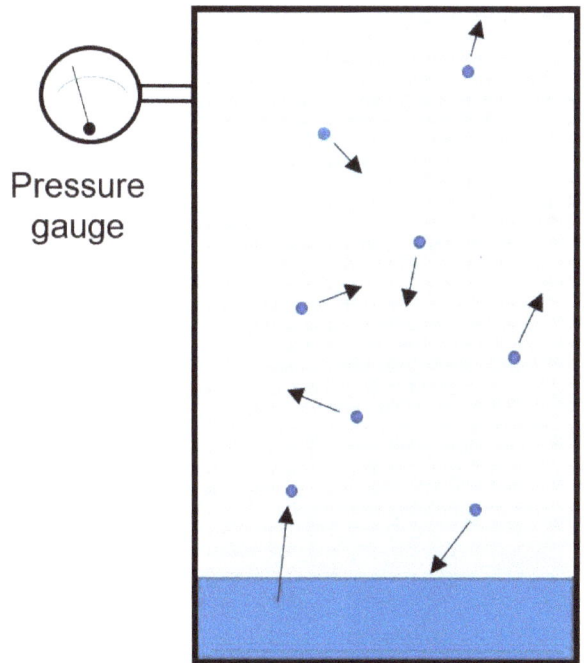

This is a "dynamic equilibrium", as molecules are continually escaping the liquid and other molecules are continually being recaptured. Any particular molecule might find itself in the liquid at 9.00 am, in the vapour at 9.45, back in the liquid at 10.15, etc.

The interesting thing is that for a given liquid (in this case, pure water) the equilibrium vapour pressure **depends only upon the temperature**. For example:

- If we increase the surface area of the liquid, by adding more water to the container or changing its shape, this increases the rate at which molecules escape from the liquid, but also increases the rate at which water molecules in the gas are recaptured back into the liquid. Consequently, the equilibrium vapour pressure doesn't change if the surface area of the liquid increases.

- If we used a vacuum pump to remove water vapour, this would initially cause the total gas pressure to drop slightly below the equilibrium vapour pressure of the water. This reduces the rate at which water molecules return to the liquid from the vapour state, while water molecules continue to escape at the same rate. The liquid water would boil to restore the equilibrium vapour pressure.

  As we continue to pump away water vapour, and as the water boils to restore the vapour pressure, vaporisation of the liquid absorbs heat and its temperature falls. The liquid becomes colder and its vapour pressure reduces. Eventually all of the liquid water would boil away, and the pressure could then fall to zero.

Note that the equilibrium vapour pressure varies dramatically – exponentially - with temperature. For water, increasing the temperature by 10°C causes the equilibrium vapour pressure to roughly double. What that means is that increasing the temperature by 20°C causes the equilibrium vapour pressure to increase about 4 times, increasing the temperature by 30°C causes the equilibrium vapour pressure to increase by roughly 8 times as much, and increasing the temperature by 40°C causes the equilibrium vapour pressure to increase by about 16 times. Over the temperature range from 0°C (the freezing point of water) to 100°C (normal boiling point), the equilibrium vapour pressure of water increases nearly two hundred times!

Equilibrium vapour pressure of water

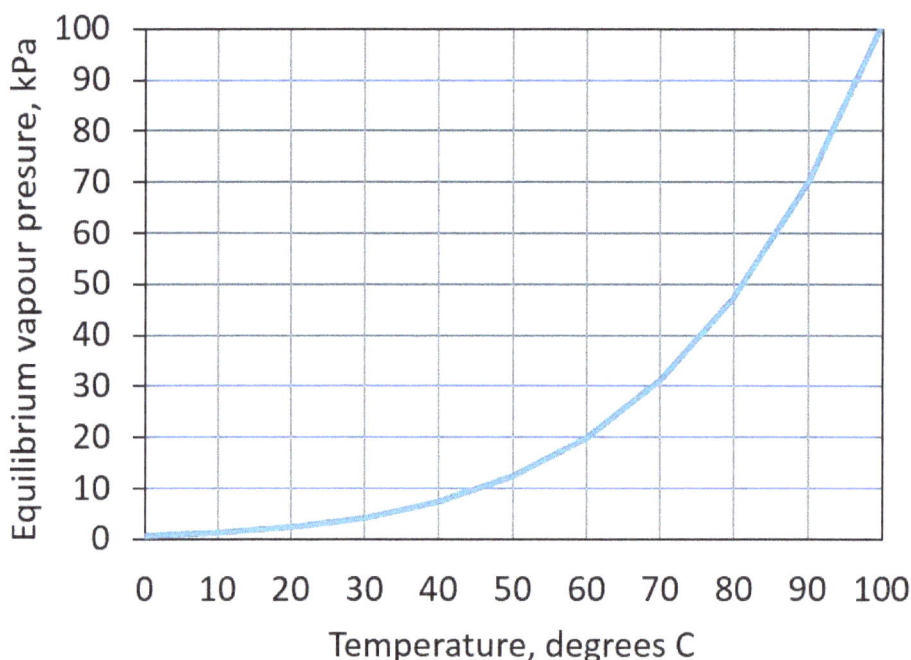

The concept of equilibrium vapour pressure is critically important for understanding natural processes occurring on Earth, and has surprising implications. For one thing, many people think that water boils at 100°C, but water can boil over a wide range of temperature, depending on the pressure. Normally, we experience air pressure of one atmosphere, at which water boils at 100°C. So, the ___normal boiling point of water___ is 100°C. However, if you lived on the top of Mount Everest, where the air pressure is only about one-third that at sea level, the boiling point of water would only be about 70°C.

The boiling point of water – or any liquid – is determined by the surrounding gas pressure. Whenever the equilibrium vapour pressure exceeds the gas pressure acting on the liquid, then any small bubbles of air or vapour will rapidly expand and rise to the surface. Rising bubbles create turbulence, which causes additional microscopic bubbles of vapour to form, and these bubbles rapidly expand and rise to the surface. Once boiling starts, evaporation is no longer limited to the upper surface of the liquid, but occurs at a huge surface area of bubbles *within* the liquid. Boiling provides a quick and effective means to remove heat by vaporising water. Heat can be provided either by a burner or other *external* heat source, or by the "sensible (*internal*) heat" of the liquid water itself.

The boiling of liquid at low pressure is shown in the following videos:

Boiling water at low pressure, Richard Hammond, 2-1/2 minutes
www.youtube.com/watch?v=XoOQNwcrDWE

Boiling water in a syringe (no need for a vacuum pump!), 4 minutes
www.youtube.com/watch?v=I5mkf066p-U

If we place water at room temperature in a vacuum chamber and then begin pumping out the air, the air pressure would decrease until the liquid begins to boil. As water molecules escape into the gas (which is then pumped away), this absorbs heat, and the liquid cools. The equilibrium vapour pressure falls as the liquid cools, so the vacuum pump must reduce the air pressure even further to keep the liquid boiling. The liquid continues to boil, gets colder and colder, until eventually the temperature of the water reaches 0°C and the water freezes. Even after water freezes into ice, water molecules continue to "evaporate" from the frozen surface, a process called "sublimation" [Note 1].

Causing liquids to boil by reducing the pressure may sound extremely bizarre and incredible, but when you see it with your own eyes many times, it begins to make perfect sense. This behaviour is not limited to water: exactly the same applies to any liquid. When I was working in an organic chemistry laboratory many years ago, I was routinely working with dilute solutions of chemical products dissolved in a liquid solvent (solvents such as acetone, ether or toluene are often used to dissolve organic compounds). To remove the solvent, leaving crystals of the pure chemical product, chemists use a piece of equipment called a "rotary evaporator" (or "rotovap"). This evaporates the solvent by applying a vacuum (at the same time, rotating the flask of liquid, so that the solution doesn't suddenly begin to boil violently and fly out of the flask). The chemistry lab where I worked was not air-conditioned, and on warm humid summer days, I would notice droplets of water condensing and dripping from the outside of the cold flask. As the solvent continued to boil, moisture on the outside of the flask would freeze into ice crystals. I was intrigued the first time that I saw this, but soon accepted it as perfectly "normal" behaviour for liquids. And, of course, it is.

You can see this yourself by watching the following short video:

Boiling/freezing acetone in a flask, 5 minutes
www.youtube.com/watch?v=oSMiec0bECw

Any liquid can be made to boil by reducing the surrounding air pressure. The "normal boiling point" of a liquid is the temperature at which it boils at one atmosphere pressure. At one atmosphere pressure, the density of water vapour is roughly one-thousand times less than the density of liquid water and, as it turns out, this is true for most liquids and their vapours. We can use this "rule-of-thumb" to explain much about why our world is the way it is.

Imagine that we could take a "snapshot" showing individual molecules in the liquid and vapour at a particular instant of time. Let's say that there are **N** molecules, within a layer one molecule thick (called a "monolayer") at the surface of the liquid. The fraction of these molecules that have sufficient energy to escape (more than the heat of vaporisation **E$_{vap}$**) is $e^{-Evap/RT}$. The number of molecules escaping from the liquid at any instant is **N**$e^{-Evap/RT}$.

That tells us half the story – the rate at which molecules can escape from the liquid to the vapour. The other half of the story is the "capture" of vapour molecules at the surface of the liquid.

Remember that, at atmospheric pressure, molecules in a gas are free to move, but frequently collide with neighbouring molecules. The situation of a molecule in water vapour is similar to the situation in which we might find ourselves on a crowded railway platform. We are free to gradually work our way to anywhere we want to be on the platform, but in the short term, we are constrained to move along with the crowd.

Molecules in the vapour are uniformly distributed in space so that, at any one time, some molecules find themselves directly above the liquid surface – within a layer about one molecule thick. Molecules so close to the liquid surface "feel" attractive forces to water molecules within the liquid below, and are pulled into the liquid.

At the normal boiling point of a liquid **T$_b$**, the equilibrium vapour pressure is one atmosphere. The density of the vapour (at one atmosphere pressure) is roughly one-thousand times less than the density of the liquid, so the number of molecules in this thin layer of vapour is one-thousand times less than the **N** molecules within the monolayer at the surface of the liquid.

At equilibrium, the rate at which molecules escape from the surface monolayer of liquid is equal to the rate at which molecules are captured from the monolayer-thick layer of vapour immediately above the liquid surface.

Consequently, at any instant in time:

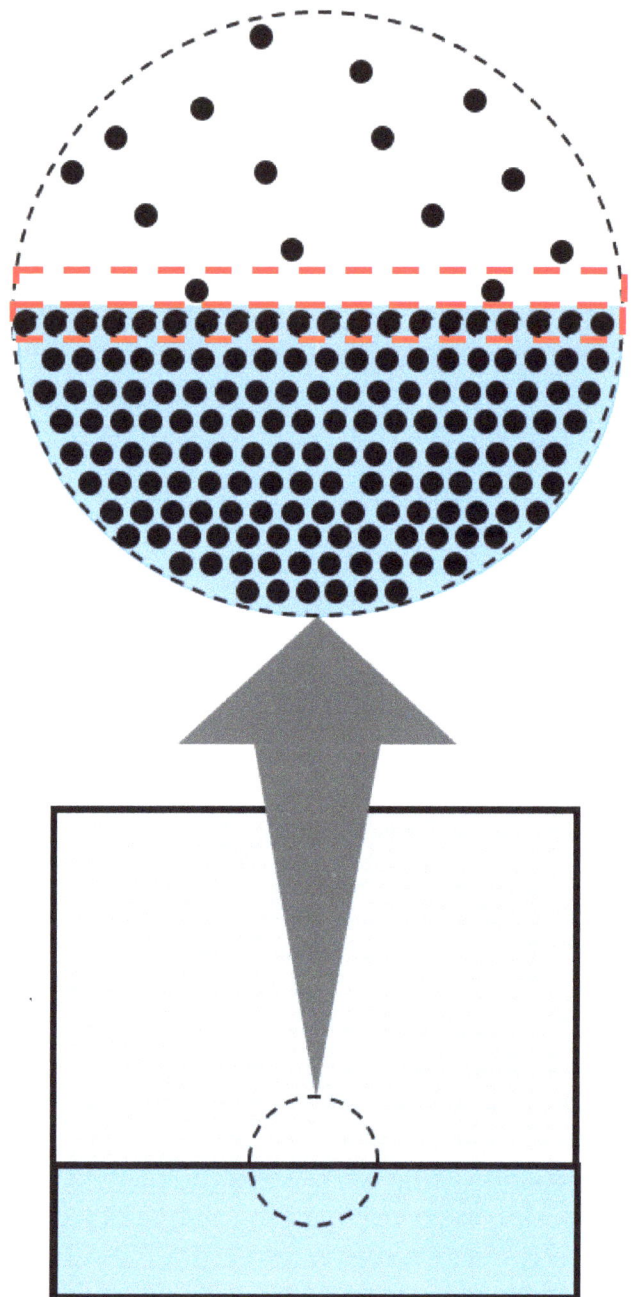

$$N\,e^{-Evap/RT_b} = \text{approximately } N\left[\frac{1}{1{,}000}\right]$$

$\underbrace{\phantom{N\,e^{-Evap/RT_b}}}$ $\underbrace{\phantom{\text{approximately } N\left[\frac{1}{1{,}000}\right]}}$

| Number of molecules escaping the liquid | Number of molecules in the vapour being recaptured |

N appears on both sides of the equation, and can be cancelled out. This leaves:

$$e^{-Evap/RT_b} = \frac{1}{1{,}000}$$

Re-arranging, we get:

$$-\frac{E_{vap}}{T_b} = \ln\left[\frac{1}{1{,}000}\right] \qquad \text{So,} \quad \frac{E_{vap}}{RT_b} = \text{approx. } 6.9$$

Since **R** is a constant (8.3 Joules/mole-deg), this gives simply:

$$\frac{E_{vap}}{T_b} = \text{approx. } 60$$

The molar heat of vaporisation E_{vap} varies in direct proportion with the normal boiling point T_b . . . . *for all liquids*! The molar heat of vaporisation is roughly 60 times the normal boiling point temperature (in degrees Kelvin). For example, butane has a normal boiling point of 0°C (273°K), so we would expect its heat of vaporisation to be about 60 X 273 = 16,400 Joules. Its actual heat of vaporisation is 21,000 Joules/mole.

This relationship is known as Trouton's rule. It allows us to estimate the heat of vaporisation for any liquid if we know its normal boiling point. Trouton's Rule only gives approximate values [see Note 2], but I think it is amazing that it gives answers within the correct ballpark for substances ranging from liquid helium (boiling point 4.2°K) to molten tungsten (boiling point 5,800°K).

I have plotted normal boiling points and heats of vaporisation as points on a graph for a wide range of substances, including "noble gases" (helium, neon, argon), organic compounds (methane, ethanol, acetone, butane) and metals (hydrogen, mercury, sodium, potassium, magnesium, thallium, lead, manganese, aluminium, tin, iron, vanadium, lanthanum, rhodium, technetium, zirconium, molybdenum, niobium, tantalum and tungsten), and other elements and compounds (water, ammonia, oxygen and nitrogen). The boiling points of these substances vary over a thousand-fold range – but, as you can see, the graph points lie close to a straight line passing through the origin (at a boiling point of absolute zero).

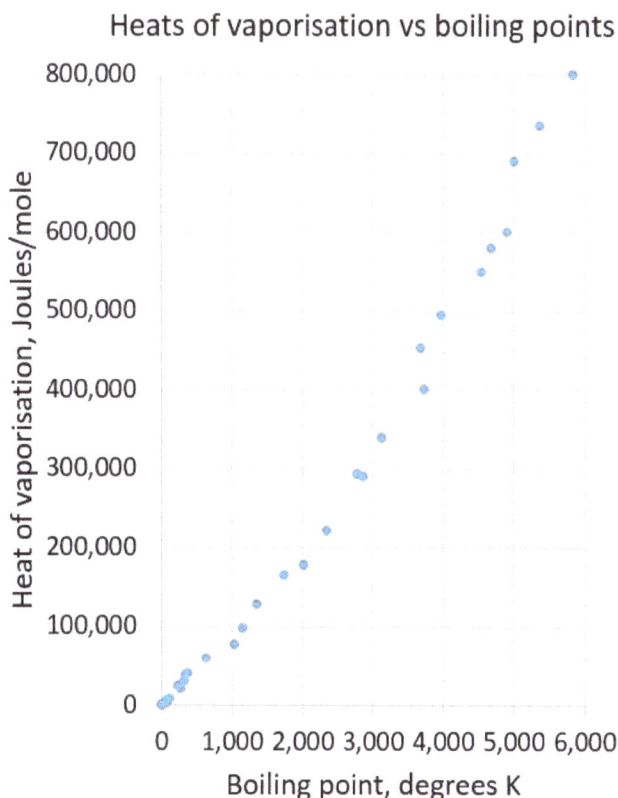

Heats of vaporisation vs boiling points

(y-axis: Heat of vaporisation, Joules/mole; x-axis: Boiling point, degrees K)

How Trouton's Rule explains our world

The chemical compounds comprising our atmosphere (mainly nitrogen and oxygen, but also argon and carbon dioxide) are considered "gases" because their normal boiling points are well below the temperatures normally encountered at the surface of the Earth. The same goes for other common "gases", such as hydrogen, helium, methane, ethane, neon or sulfur dioxide. However, if these compounds are present on distant planets, in the frozen outer reaches of their solar systems, such compounds could form liquid oceans, or icy glaciers or solid rocky crust. In fact, the planets Uranus and Neptune, orbiting in the distant outer reaches of our solar system, have "rocky" surfaces composed of frozen water, ammonia and methane ice. Their moons are believed to have solid surfaces composed of frozen water or nitrogen.

Compounds are considered "gases" in our world because intermolecular forces holding the molecules together are quite weak, and therefore, their Molar Heat of Vaporisation is relatively small (in relation to the available heat energy **RT**). At the temperatures encountered on Earth, the intermolecular forces are too weak to contain these molecules within the liquid state. But, at very low temperatures, the molecules succumb to intermolecular forces and are trapped in the liquid state or bound into solid crystals.

Substances that we consider as "liquids" (like water, kerosene, alcohols or acetone) have boiling points above "room temperature". These the molecules are held together by stronger intermolecular forces, and their Heat of Vaporisation is greater than in compounds that we consider to be "gases". Nonetheless, at higher temperatures (or lower pressures), these compounds exist as gases.

Finally, those substances that we consider to be "solids" (like steel, aluminium and rocks) are held together by very strong intermolecular forces between their constituent atoms or molecules. Accordingly, these materials have high melting points and very high boiling points, and their Molar Heats of Vaporisation are much greater than the compounds that we normally consider to be "gases" or "liquids".

Notes

1. Removal of water by applying vacuum to frozen solutions is the basis of freeze drying, used to produce instant coffee and other food products. The advantage of freeze drying is that water is removed at sub-freezing temperatures, so the process doesn't affect complex, heat-sensitive molecules that give foods their characteristic flavours and aromas.

2. One liquid which doesn't conform closely with Trouton's Rule is water. With its normal boiling point of 373°K, we would expect the Heat of Vaporisation of water to be about 22,000 Joules/mole. This is about half the actual Heat of vaporisation of water.

8. Keeping cool:
The science of refrigeration and air-conditioning

Perhaps no invention has played a greater role in transforming our society and economy than the refrigerator. Until the advent of large industrial refrigeration plants in the mid-19th century, fresh food could only be transported and distributed over short distances and times. Production of fruit, vegetables, dairy products and meat had to be undertaken within short distances of towns and cities. But the nexus between agricultural production and population centres was broken by refrigeration and freezing. It enabled the agricultural sector and cities to develop largely independently, and catalysed the amalgamation of small family farms into huge, highly mechanised operations, with a massive shift of population from rural to urban areas.

This was perhaps the greatest social revolution of modern history, and refrigeration played a major part, especially in the latter stages. Only several hundred years ago, the vast majority of people lived on farms or in small agricultural communities. During the early history of the United States, about 95% of the population was involved in farming, and 5% lived in towns and cities. Within three hundred years, this completely reversed. This trend is still continuing in Asia and elsewhere as family-owned farms are amalgamated and mechanised, and as rural populations shift into cities.

The technology used in modern refrigerators was invented by James Harrison, a Scottish-born inventor living in Geelong, Victoria. The story of James Harrison and how he came to invent a commercial refrigeration system was told in a recent television documentary "Great Aussie Inventions that changed the world".

James Harrison was the founder and editor of the Geelong Advertiser newspaper, but appreciated the huge need for ice (to preserve food and cool drinks) that could not adequately be met by importing huge ice blocks from overseas. Harrison had seen ether used to clean metal type used in printing newspapers, and he noticed that the ether cooled the type as it evaporated. He developed a steam-powered refrigeration system which used ether as a refrigerant to produce chilled brine (which he then used to make ice).

Harrison patented this technology and established the first commercial ice-making machine in 1854. Based on successful ice-making operations in Geelong and Sydney, Harrison proposed to use his refrigeration technology to freeze meat and export it to Britain in insulated containers. He persuaded investors to fund a trial shipment but, unfortunately, the insulated containers were inadequate to keep the meat frozen. The entire shipment of meat spoiled on the journey, which was a financial disaster and undermined public confidence in refrigeration. Harrison set aside his development of refrigeration technology and returned to journalism, becoming Editor of the Melbourne Age. Subsequently, the refrigeration technology invented by Harrison was refined by other entrepreneurs used to build large commercial refrigeration and freezing plants in the United States and other countries.

The introduction of the first household refrigerators, starting around 1914 in the United States, would ultimately change the way people lived in cities and surrounding suburbs. No longer would people need to shop each day at their local baker, butcher and other small shopkeepers. As late as 1960, as I was growing up in the Lower East Side of New York City, I recall a bustling world on Avenue B of small shopkeepers, bakeries, butcher shops, a delicatessen and wizened-looking ladies and men selling products from carts or pickles from a barrel (although there was also a small supermarket on Avenue D). But then, with most families owning a refrigerator and growing numbers owning a car, householders (usually housewives) starting to do weekly grocery shopping at supermarkets and shopping plazas. This, combined with many other factors, enabled women to join the workforce and changed the social fabric of society to what it is today – in many ways for the better.

So how do refrigerators work?

Operation of refrigeration and air-conditioning systems is based on the evaporation/boiling of volatile liquids (refrigerants). As we have seen in the previous chapter, a liquid can be made to boil at nearly *any* temperature by reducing the pressure. Furthermore, we have seen that the process of evaporation (or boiling) absorbs heat. If we cause a liquid to boil by reducing its pressure, the heat of vaporisation is provided by internal heat within the liquid itself, causing its temperature to fall.

Using these simple concepts, let's see how we might design a refrigerator. First, we need to choose a suitable liquid that we could boil by applying a vacuum. We could use water. It is cheap, non-toxic, non-flammable and widely available, but it has two major shortcomings as a refrigerant. Firstly, at the temperature within a refrigerator (0-6°C), the equilibrium vapour pressure of water is only about 1/100th atmospheric pressure. We would need a very large pump to suck sufficient volumes of water vapour at such low pressure. Secondly, to maintain temperatures with a fridge at just above 0°C, the refrigerant would need to be even colder (to absorb heat entering the fridge), so a water refrigerant would freeze. And, of course, water would be unsuitable as a refrigerant at sub-freezing temperatures.

The ideal refrigerant would be more volatile than water. It would have a normal boiling point at, or below, normal room temperature and would have a very low freezing point. Of course, the ideal refrigerant should also be cheap, non-toxic, non-flammable, non-corrosive, environmentally benign and readily available. Unfortunately, no known refrigerant fully meets all these requirements.

Early industrial refrigeration plants used liquid ammonia as refrigerant. With a normal boiling point of -33°C, ammonia has sufficiently high vapour pressure and other properties to be a practical refrigerant. However, ammonia is toxic at high concentrations, and fatal accidents involving leaks at commercial refrigeration plants caused ammonia to fall out of favour. Other compounds used as refrigerants, such as methyl chloride and sulfur dioxide, were also highly toxic.

In the 1950s and 1960s, a series of synthetic chemical compounds called "freons" became widely adopted for refrigeration. The most common was Freon-12 (dichlorodifluoromethane, with a normal boiling point of -30°C). These compounds seemed ideally suited as refrigerants. They are extremely chemically stable, and so, don't burn and don't decompose in the atmosphere. Years later, researchers discovered that these compounds eventually diffuse to the upper atmosphere, where they release chlorine free radicals which destroy ozone molecules that provide a protective shield that absorb ultraviolet light in sunlight. The ozone layer in the upper atmosphere performs a critical role protecting humans, as well as all other animals and plants living on the surface of the Earth. Once the threat of ozone destruction was recognised, the production and use of freon compounds was phased out and prohibited through the 1987 Montreal Protocol. This was one of the very few highly effective international collaborations to protect the environment.

The class of compounds that was adopted to provide a direct substitute for freons in the short-to-medium term was fluorinated hydrocarbons, such as HFC 134a (1, 1 ,1 ,2 – tetrafluoroethane, normal boiling point = -26°C). These compounds have a very low ozone depletion effect, but are very powerful greenhouse gases. For example, HFC 134a has a "Global Warming Potential" that is 1,430 times that of carbon dioxide.

Eventually, fluorinated hydrocarbons like HFC 134a are expected to be phased out. Some refrigeration and air-conditioning systems use hydrocarbons, such as propane (normal boiling point of -42°C) or butane (normal boiling point of -1°C). Hydrocarbon compounds pose a fire risk in the event of a leak, although the risk can be minimised by incorporating relatively small quantities of refrigerant within the system. Eventually, most refrigerators and air-conditioners are likely to use hydrocarbon refrigerants.

Let's say that we choose pentane as a refrigerant. Pentane is a liquid at room temperature but, with a normal boiling point of 36°C, it readily boils when the pressure is reduced just below atmospheric pressure. The freezing point of pentane is -130°C, which is well below the temperatures produced in a refrigerator or air-conditioner, so freezing is not an issue.

To make a refrigerator, we could simply mount a container of pentane within an insulated refrigerator cabinet, and use a vacuum pump to reduce the pressure of pentane vapour, causing it to boil. As the pentane boils, it absorbs heat and chills the vegetables, cheese and milk inside the refrigerator cabinet. The container of boiling refrigerant is called an "evaporator". It is normally made of metal tubing and has fins to increase the surface area for heat transfer. You probably won't see the evaporator coil inside your home fridge, as it is usually hidden behind a plastic cover with ventilation slots (the metal fins are easily damaged). Usually, a small fan circulates cold air over the evaporator coil and around inside the fridge.

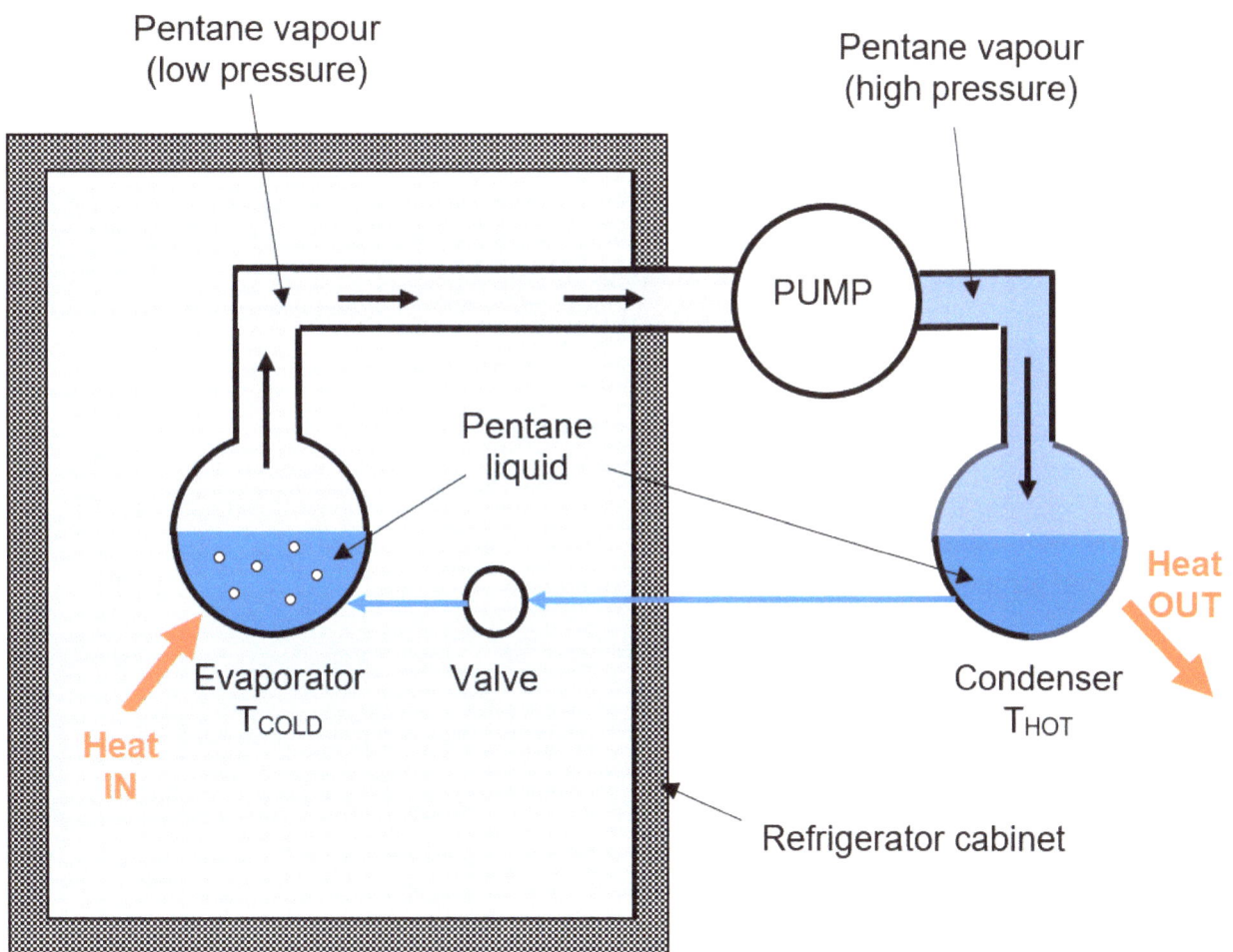

Pentane vapour (low pressure)

Pentane vapour (high pressure)

PUMP

Pentane liquid

Heat OUT

Evaporator T_{COLD}

Valve

Condenser T_{HOT}

Heat IN

Refrigerator cabinet

The boiling pentane absorbs heat, cooling the contents of the fridge, but this is only half the story. What happens to pentane vapour that is sucked from the evaporator? We don't want to discard this into the surrounding atmosphere, and then have to continually add more liquid pentane to replace the liquid that boils away.

The pentane vapour is pumped into another container mounted **outside** the refrigerator cabinet. The vapour is forced to condense back into liquid by increasing its pressure above the equilibrium vapour pressure. Just as we can cause a liquid to boil at low temperature by **reducing** the pressure, we can cause vapour to condense into liquid by **increasing** its pressure. The container used to condense refrigerant vapour is also commonly made of metal tubes with fins, and (as you might expect) is called a "condenser". The condenser coil is commonly mounted at the rear of the fridge (and, in recent models, is often hidden inside the back panel).

Eventually, liquid pentane within the evaporator would boil away, while liquid pentane would collect in the condenser. To avoid this, liquid refrigerant is continually recycled from the condenser to the evaporator through an automatic control valve. No pump is needed to recycle the liquid, since the condenser is at higher pressure than the evaporator.

The operation of a typical refrigerator is described in this 4-1/2 minute video:
https://www.youtube.com/watch?v=qrcEWhurQl4

Energy required for refrigeration

A refrigerator or air-conditioner is a "Heat pump". It transfer heat from a "Heat Source" (inside a cold fridge) to a "Heat Sink" (the surrounding air at higher temperature). The "heat source" is at lower temperature than the "heat sink" so, effectively, a refrigerator pumps heat "uphill". To do so, it must consume energy.

The Second Law of Thermodynamics states that heat cannot be transferred from a cooler to a hotter body **unless work is done**. We all know this intuitively. We have all seen that a hot cup of tea will cool to room temperature all by itself. We also know that a warm bottle of beer will not get cold if we leave it on the kitchen counter. If we want to chill a warm bottle of beer, we must do work – or rather, the pump inside our refrigerator must do work. So, how much work is required to transfer each Joule of heat energy?

Let's assume that we have built the "perfect" refrigerator – one that has no friction, no losses, and perfect heat transfer from the evaporator and condenser. In fact, this would be the most efficient refrigerator that is allowed by the Second Law of Thermodynamics. In our "perfect fridge", how much work must be done to transfer an amount of heat Q_{COLD} from inside the refrigerator at temperature T_{COLD} to the surrounding air at temperature T_{HOT}.

The heart of the refrigerator is the pump, which (literally) does all the work. Basically, the pump causes pentane liquid to boil at low temperature T_{COLD} (say, 5°C inside the fridge), and causes pentane vapour to condense back to liquid at higher temperature T_{HOT} (say, 25°C within our kitchen).

Since the condenser is mounted outside the refrigerator, where the temperature is higher, the pressure inside the condenser is higher than in the evaporator.

58

Vapour pressure at T_{COLD} and T_{HOT}

The pump must do work in sucking pentane vapour from the cold evaporator, where the vapour pressure P_{COLD} is low, and then compressing and pushing the vapour into the hot condenser, where the pressure P_{HOT} is high. In other words, *the pump must work against a pressure difference resulting from the difference in the equilibrium vapour pressure of the refrigerant at T_{COLD} and T_{HOT}*

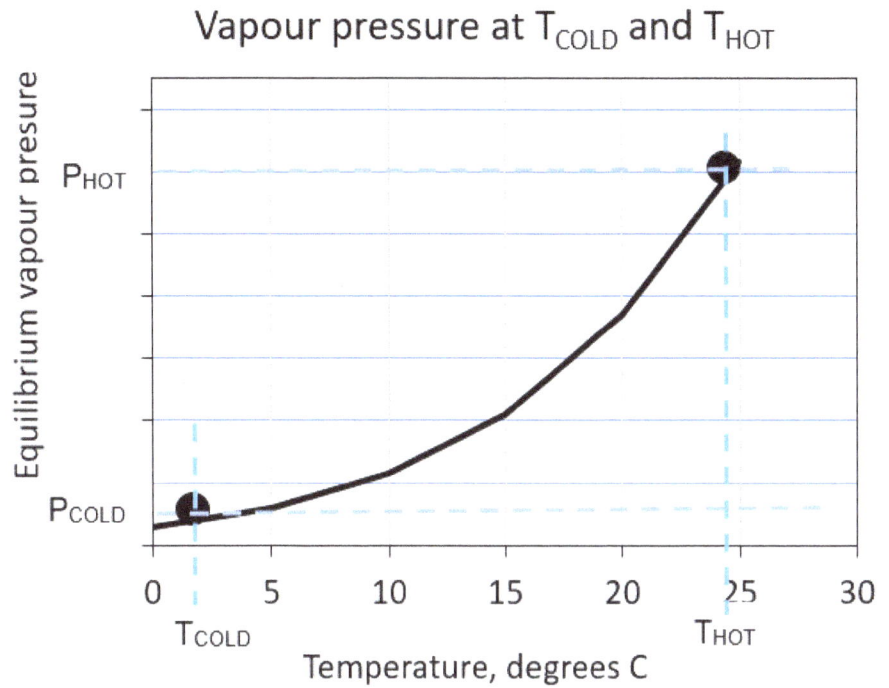

Pentane refrigerant cycles continually from liquid to vapour (inside the fridge at low temperature and low pressure), and then from vapour back to liquid (at high temperature and pressure). For this to occur, the pump must compress low-pressure refrigerant vapour at P_{COLD} to high pressure P_{HOT}.

We have seen that the vapour pressure of any liquid varies exponentially with temperature, so the *ratio* of pressures P_{HOT} and P_{COLD} also varies exponentially with temperature as follows:

$$\frac{P_{HOT}}{P_{COLD}} = \frac{e^{-Evap/RThot}}{e^{-Evap/RTcold}} = e^{-Evap/R[\frac{1}{Thot} - \frac{1}{Tcold}]}$$

It is convenient to express this equation in logarithm form, as:

Eq (1)
$$Ln\left[\frac{P_{HOT}}{P_{COLD}}\right] = -\frac{Evap}{R}\left[\frac{1}{T_{HOT}} - \frac{1}{T_{COLD}}\right]$$

We have also seen that the work required to compress a gas varies directly with the ratio of the final and initial pressures. In this case, the gas is initially at pressure P_{COLD} and is compressed to a final pressure of P_{HOT}:

Eq (2)
$$\text{Work required to compress a gas} = n\,R\,T_{HOT}\,Ln\left[\frac{P_{HOT}}{P_{COLD}}\right]$$

Combining equations (1) and (2) tells us how much work is expended in taking **n** moles of refrigerant gas from the cold evaporator (at temperature T_{COLD}), compressing it, and pushing it into the condenser (at temperature T_{COLD}):

$$\text{Work required to compress gas} = n\,\cancel{R}\,T_{COLD}\left[-\frac{Evap}{\cancel{R}}\right]\left[\frac{1}{T_{HOT}} - \frac{1}{T_{COLD}}\right]$$

In calculating the work required to compress the refrigerant gas, I have assumed that the gas is compressed at constant temperature. Technically speaking, the gas is not actually compressed at constant temperature - but as it turns out, this doesn't matter. For most refrigeration and air-conditioning systems, the temperature difference between the heat source and heat sink ($T_{HOT} - T_{COLD}$) is small -- much less than the absolute temperature T_{COLD}. So, in fact, there is hardly any difference to the work required whether or not the gas absorbs heat from its environment.[Note 1]

Even if the temperature difference between T_{COLD} and T_{HOT} is large, it is possible to transfer heat through 2, 3 or (at least in principle) *many* refrigeration stages. Then, each stage operates with a small temperature difference, and the pump only needs to compress the refrigerant vapour to slightly higher pressure, so the gas is compressed at nearly constant temperature.

Two-stage refrigeration

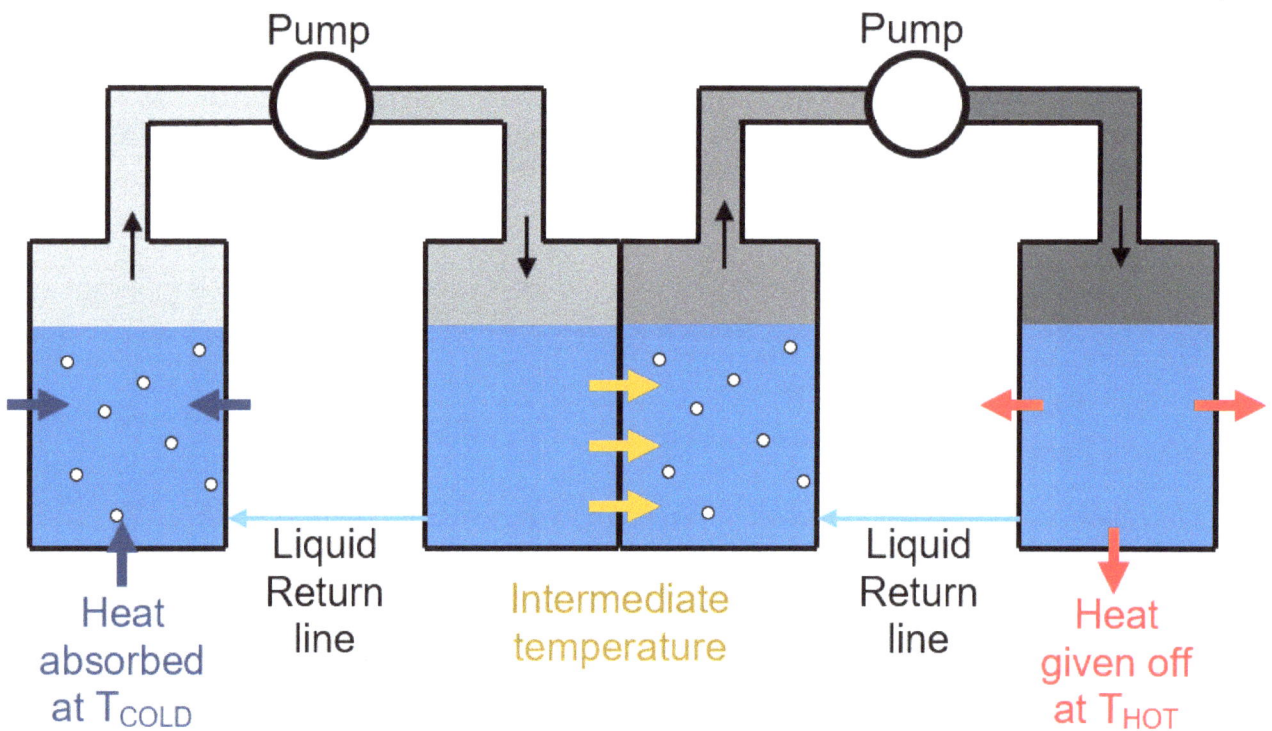

For a refrigerator or air-conditioner, we want to *transfer the maximum amount of heat* from the heat source with the *minimum work input*. We define the "Coefficient of Performance" (COP) of a refrigerator or heat pump as the ratio of the heat energy removed divided by the work input. From the point of view of energy efficiency, we want the Coefficient of Performance to be as high as possible so that we need do the minimum work (use the minimum energy) to chill our beer and keep our vegetables cold.

Each mole of refrigerant liquid that evaporates absorbs its heat of vaporisation E_{vap}. So. the heat absorbed by evaporating **n** moles of liquid refrigerant is $n\,E_{vap}$.

The "Coefficient of Performance" is the ratio of the heat absorbed to the work input required.

$$\text{Coefficient of Performance} = \frac{\cancel{n}\, E_{vap}}{\cancel{n}\, T_{COLD}\, E_{vap}\left(\frac{1}{Tcold} - \frac{1}{Thot}\right)}$$

There is little difference in absolute temperature between T_{HOT} and T_{COLD}, so this equation can be simplified to an extraordinarily simple result, which states the maximum possible Coefficient of Performance of a refrigerator or other "heat pump":

Eq (3) $$\text{Maximum Coefficient of Performance} = \left[\frac{T_{COLD}}{T_{HOT} - T_{COLD}}\right]$$

The minimum work required to remove heat Q_{COLD} from the refrigerator (or inside our house) and transfer it to the outside environment is:

Eq (4) $$\text{Minimum Work required} = Q_{COLD}\left[\frac{T_{HOT} - T_{COLD}}{T_{COLD}}\right]$$

Imagine that we have a perfect refrigerator/freezer - one that is as efficient as it could possibly be. Let's say that we buy a kilogram of steak at the supermarket, bring it home, and put it into our perfect freezer (which, by Australian standards, must maintain its contents at -18°C). Steak, like many foods, consists mostly of water. As the steak first cools down to freezing temperature, then freezes, and then is chilled further to -18°C, it will release 400,000 Joules of heat. At a room temperature T_{HOT} of 30°C, and with T_{COLD} being -18°C (255° absolute), the maximum coefficient-of-performance of the freezer is about 5. In chilling the steak to -18°C, the freezer consumes at least 80,000 Joules of energy, or 0.02 kilowatt-hour of electricity (costing less than one cent).

Bear in mind that the COP of a refrigerator or an air-conditioner only tells part of the story. Even a refrigerator with a high COP would consume lots of energy if poorly insulated walls or leaking door seals allows lot of heat to enter the refrigerator cabinet.

Electricity consumption by a mid-size fridge/freezer would typically cost about $80-100 per year. Over its lifetime, electricity used by a typical fridge would cost more than its initial purchase price.

The same story applies to air-conditioners. Householders buying an air-conditioner should choose a model with a high COP (as indicated by the star rating) to minimise their power consumption and electricity bills. But, again, that is only part of the story. Even an efficient air-conditioner would use excessive energy if heat enters the building through unshaded windows and uninsulated roof and walls.

If we are air-conditioning a house, where the inside temperature is 23°C (296° absolute) and the outside temperature is 10°C hotter than inside, then it would theoretically be possible to achieve a COP as high as 30. In reality, current air-conditioners do not achieve a COP nearly that high. A typical room air-conditioner might achieve a COP of 3-4. A listing of actual measured COPs for all air-conditioner models sold in Australia is available on the Energy Rating website operated by the Australian Government (www.energyrating.gov.au).

But even a COP of 3 is still a bargain. We only need to put in one unit of work energy to remove 3 units of unwanted heat energy. It looks like we are cheating the Laws of Thermodynamics, but we are not. A refrigerator does not get rid of heat. It takes heat from inside the refrigerator cabinet (at about 2°C, or 275° absolute) and effectively pushes the heat "uphill" into the surrounding environment (at around 25°C, or 298° absolute).

The consequences of the 2nd Law of Thermodynamics are clear: To achieve the highest efficiency (COP) for any heat pump, the temperature difference against which the heat pump is operating, $T_{HOT} - T_{COLD}$, should be as small as possible (relative to the absolute temperature T_{COLD}).

Other applications of heat pumps

Even in applications involving moderately large temperature differences, heat pumps can still provide significant reductions in energy consumption. For example, heat pumps provide an effective alternative to solar water heating by using the air as a heat sink. In Queensland, daytime temperatures are typically 20-30°C, so the air can readily serve as a heat source for water heating.

Water heating accounts for about 30% of the total energy used in a typical Queensland household, as well as much of the energy used in restaurants, laundries and many other commercial and industrial operations.

In the past, many households used hot water storage tanks heated by electric resistance heating. This was encouraged by the availability of cheap off-peak electricity tariffs (which allowed water to be cheaply heated during night-time hours, and stored for daytime use). Electric resistance heating is wasteful, since it converts a high-grade energy source (electricity) into low-grade energy (heat at about 60°C).

In some cases, electricity is the only practical energy source available for water heating and, in these situations, heat pump water heaters are far more efficient than conventional electric water heaters.

A heat pump water heater is basically a refrigerator – but operated under different conditions for a different purpose. It absorbs heat from the surrounding air (typically at 20°C), and "upgrades" and pumps this low-grade ambient heat into a water tank at higher temperature (typically 60°C).

The maximum possible Coefficient of Performance that could be achieved by transferring heat from air at 20°C (293° absolute) to a water tank at 60°C is given as follows:

$$\text{Maximum Coefficient of Performance} = \frac{T_{COLD}}{T_{HOT} - T_{COLD}} = \frac{293}{60 - 20} = 7.3$$

Typical heat pump water heaters have a COP of about 3, which means that they use only one-third as much electrical energy as a conventional electric water heater - while providing the same amount of hot water at the same temperature. If the source of electricity is coal-fired power stations, only one-third as much coal must be burned, releasing one-third the amount of greenhouse gases and other pollution (nitrogen oxides, sulphur oxides, particulates, etc). That's a huge advantage from an environmental viewpoint, and a huge benefit to the local and global community.

The operation of a heat pump water heater is depicted in this one-minute video:
https://www.youtube.com/watch?v=sxDfdI1IHnE

From an economic viewpoint, the advantage of heat pump water heaters to the end user is not so clear-cut. Firstly, heat pump water heaters are significantly more expensive than conventional electric resistance water heaters. Consumers tend to choose the option with the lowest upfront cost, even if this option will cost more over a typical 15-year lifetime of a water heater. To be economically competitive, energy cost savings during the lifetime of a heat pump water heater should offset the higher purchase cost of the unit. Secondly, many heat pump water heaters do not have sufficient capacity to operate only during night-time hours, and therefore, cannot take advantage of cheap off-peak electricity tariffs (available only at night).

So, depending upon electricity tariffs and other factors, the "life-cycle cost" of heat pump water heaters tend to be roughly comparable to those of conventional electric water heaters, even though they use much less electricity. Until recent years, there was no strong **economic** incentive for a householder to choose a heat pump water heater instead of an electric resistance water heater. This represented a "market failure", or a dis-connect between what is in the best interests of the community (human society) and the economic interests of the buyer. Within the past decade, however, higher electricity tariffs and government regulations have shifted the balance away from conventional electric water heaters.

Heat pumps have also been applied to clothes driers, another type of energy-intensive appliance. Conventional clothes driers use electric resistance heaters to heat air circulating through the drier. The heated air evaporates water from the clothes, and the humid air is then vented into the laundry (or outside the house, through a hose). In these conventional driers, the electrical energy input provides the 2.2 megajoules of heat energy needed to vaporise each litre of water.

Heat pump clothes operate differently. Once again, hot, dry air passes over the wet clothes to evaporate water – but the warm humid air serves a the "heat source". It loses heat to refrigerant liquid that boils inside evaporator coils. As the humid air cools, water vapour condenses back into liquid, which collects within a reservoir (to be removed later). Then, the same air is re-heated by passing over a second coil in which the refrigerant vapour is condensed. Because the same air recirculates again and again, and water evaporated from the wet clothes is condensed and collected, heat pump driers are particularly suitable for indoor laundries with limited ventilation (as may be the case in high-rise apartments).

Heat pump driers have become available for sale in Australia in recent years, and comprise about half the model clothes driers available. The purchase price of heat pump driers is about twice that of conventional electric driers, but they use half the electricity. The energy savings of an 8-kilogram heat pump clothes drier (under typical conditions of use) would amount to about 150 kilowatt-hours per year. At an electricity tariff of $0.25/kWh, the savings in electricity should repay the additional purchase price within the lifetime of the drier.

Here is a 2-1/2 minute video, made by a manufacturer, explaining the operation of its heat pump driers:
https://www.youtube.com/watch?v=IWqNmOLHwt0   (O = capital letter, 0 = zero)

In Australia, some people use clothes drying technology which is even more efficient than heat pump driers – hanging clothes on a line (such as the iconic Hills Hoist), which relies entirely on renewable solar energy. Of course, this venerable technology has its own limitations, which are becoming more pronounced as people increasingly live in high-rise units and small house allotments, and increasingly expect to clean/dry their laundry whenever they feel like it (including rainy/cloudy days).

The case of heat pumps illustrates some important general points relating to energy-efficiency. Technology is obviously a critical part of the solution to reduce the world's energy and greenhouse challenges, but it is only one component. More efficient technology has generally been adopted only if it is economically competitive, or if governments intervene in the market (and governments are loath to impose restrictions or costs on voters, when the benefits will accrue to the entire world). Furthermore, technology is only be accepted and adopted if it is consistent with people's expectations, desires and culture.

Notes

1. I have made several simplifications in calculating the Coefficient of Performance of a refrigerator. The energy required to operate a refrigerator can be calculated in a much more sophisticated way, taking into account:

   - The temperature of the refrigerant gas increases as it is compressed,

   - The pump must do more work in pushing the (hotter) compressed refrigerant gas into the condenser than the work done by the cooler gas as it enters the pump.

   - Condensed liquid returning from the condenser is at higher temperature than the evaporator. Heat carried by the returning liquid is an additional "heat load" that must be transferred by the refrigeration system.

   Taking these factors into account involves more complicated calculations, however, in the situation where the temperature difference $T_{HOT} - T_{COLD}$ is much less than the absolute temperature, the effect of all of these factors becomes insignificant.

   Since, in principle, heat can be transferred in many refrigeration stages, the temperature difference in each stage $T_{HOT} - T_{COLD}$ can always be reduced to a small value. Thus, the Coefficient of Performance of an ideally-efficient refrigerator will be given by the result calculated here.

   For a single-stage refrigeration system operating with a very large temperature difference $T_{HOT} - T_{COLD}$, the actual Coefficient of Performance will be less than the result calculated here. *The calculated result given in Equation 3 sets an upper boundary for the maximum Coefficient of Performance that is possible in an ideal refrigeration system.*

9. Heat engines and the limits of power production

The huge shift of human population from farms and farming communities into major cities and population centres within the last two centuries was enabled by refrigeration technology. Since perishable foodstuffs could now be preserved and transported large distances, the human population could concentrate in metropolitan areas with millions of people, provided with food grown hundreds or thousands of kilometres away.

The social revolution enabled by heat pump technology occurred the same time as another technological revolution – heat engines. These have become so ubiquitous that we completely take for granted the engines that literally power our economy and society. Until the first steam engines were developed in the 18th century, industry, agriculture and transport relied almost entirely on human muscle or draught animals (mainly horses and oxen).

The human body is a machine that relies on food as a source of energy. A typical adult consumes a diet containing about 8-10 million joules (mostly in fats, oils and carbohydrates). Not all of this energy input is converted into a useful work output. An elite athlete (say, a bicyclist in the Tour de France) can produce about 3 million Joules of work per day. However, few people could produce or maintain such a high level of work output on a day-to-day basis. Someone doing "hard manual work" would probably produce about one million Joules of mechanical work per day.

In modern western societies, very few people still do hard physical work. Every industry now uses powered tools or powered vehicles to do physical work, with the human operator mainly guiding the machine. These machines derive their energy by burning fossil fuels (gasoline, diesel fuel, natural gas, etc) or using electricity (which is generally produced by burning fossil fuels). The daily work output that was once produced by a hard-working labourer can now be produced by a gasoline engine using one-tenth litre of gasoline (costing about 15 cents) or 0.3 kilowatt-hours of electricity (costing about 7 cents). The fuel required to produce the same work output as a strong labourer would cost about *2,000 times less* than the laborers wages.

At the start of the industrial revolution, humans began to exploit wood, and then coal and petroleum, to meet the rapidly-growing demand for mechanical work. In particular, mechanical energy was needed for pumping water out of coal mines, spinning of textiles, and a wide range of other industrial uses. This led to the development of engines, upon which modern society is based.

The first steam engine to be applied commercially was developed by Thomas Savery in 1698. The basic design was improved by Thomas Newcombe in 1712, and then by James watt in 1765. An estimated 2,000 steam engines were in operation by 1800. These early engines were extremely simple and crude by today's standards. They were grossly inefficient, converting less than 1% of the fuel energy into mechanical work output. However, over the following century, improvements in engine design and construction greatly increased their efficiency, with 20-30% of the fuel energy being converted into work output. But engine designers began to realise that the efficiency of their heat engines was approaching a limit. Further increases in efficiency were possible, but only up to the fundamental limit imposed by the 2nd Law of Thermodynamics. In fact, the fundamental law that limits the efficiency of heat engines is exactly the same as what limits the coefficient-of-performance of refrigerators and heat pumps. This might be as one might expect, as a heat engine is a heat pump that operates in reverse.

A refrigerator or "heat pump" transfers heat from a cold "heat source" to a hot "heat sink". Effectively, a refrigerator pumps heat "uphill", and this requires a work input. We can use very similar technology to do exactly the opposite – to **produce** work by taking in heat at high temperature (T_{HOT}); allowing it to flow "downhill" to the temperature of the surrounding environment (T_{COLD}); converting some into useful work, and then discarding the remaining waste heat to the atmosphere. This is the basis of a "heat engine". In fact, an Organic Rankine Cycle engine (or a closed-cycle steam engine) has the same components as a refrigerator. It uses the same type of volatile liquid (say, pentane or butane or water) as a refrigerator, and works on exactly the same principle – but in reverse. This time, the pressure of the volatile liquid in the evaporator (now called a "boiler" in steam engine terminology) is greater than the pressure in the condenser, so the vapour does work as it expands in a piston pump or turbine.

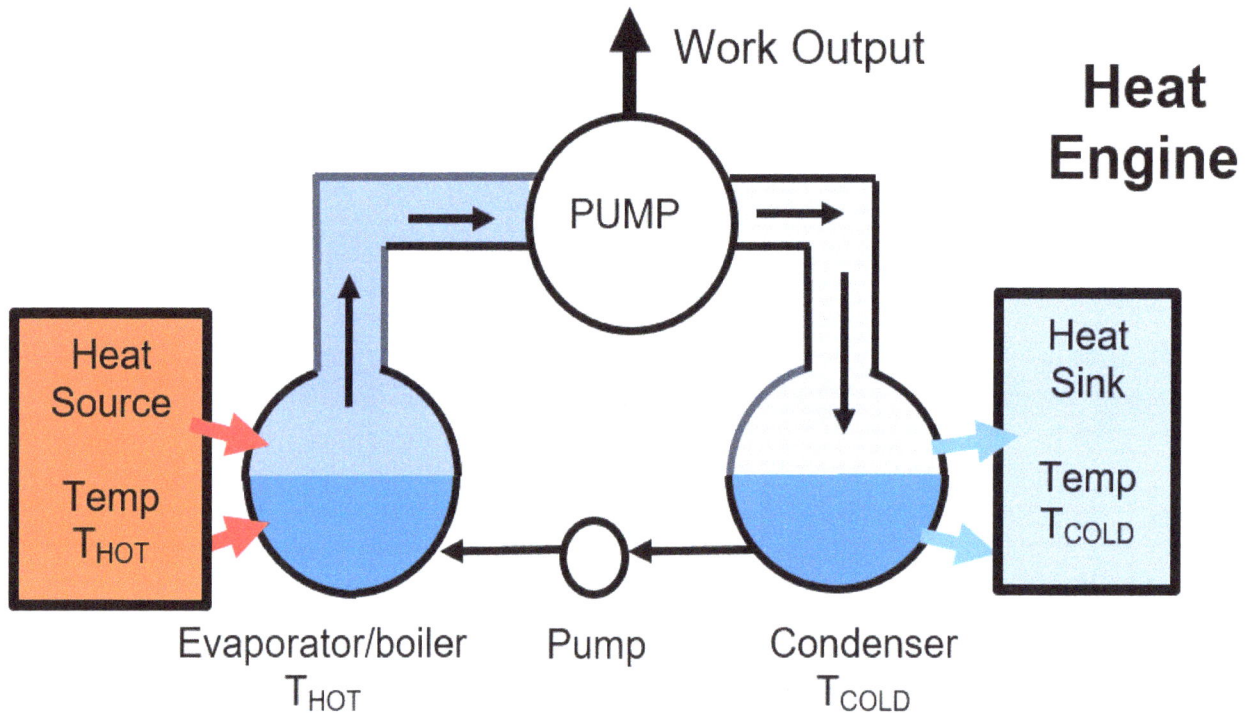

We have seen that the Second Law of Thermodynamics sets a limit for the **minimum work** required to transfer heat "uphill" from temperature T_{COLD} to high temperature T_{HOT}. In the same way, the Second Law of Thermodynamics sets a limit for the **maximum work** that can be produced when heat is transferred "downhill" from a high temperature heat source T_{HOT} to the surrounding environment at temperature T_{COLD}. Only a fraction of the heat energy can be converted into a useful work output.

We can determine the maximum efficiency of a heat engine by conducting a "thought experiment". We could, in principle, operate a "perfect" heat pump refrigerator alongside a "perfect" heat engine. The heat output of the refrigerator Q_{HOT} serves as the heat input for the engine, and the heat output of the engine is recycled as the heat input Q_{COLD} for the refrigerator. In this way, heat and work can be recycled around and around between the heat pump and the heat engine.

Imagine a situation where we have an ideal refrigerator (achieving the maximum possible Coefficient-of-Performance) transferring heat from a low-temperature heat reservoir (at T_{COLD}) to a high-temperature heat reservoir (at T_{HOT}). Then, we recycle the heat through an ideal heat engine, which takes heat from the high-temperature reservoir, converts some of the heat into work, and then discards the remaining waste heat into the low-temperature reservoir.

Heat is continually cycled from the high-temperature reservoir through the heat engine (where it produces work going "downhill") and then is dumped into the heat sink. At the same time, heat is absorbed from the low-temperature reservoir, pushed "uphill" by the heat pump (requiring work to be done), and then pushed into the heat source.

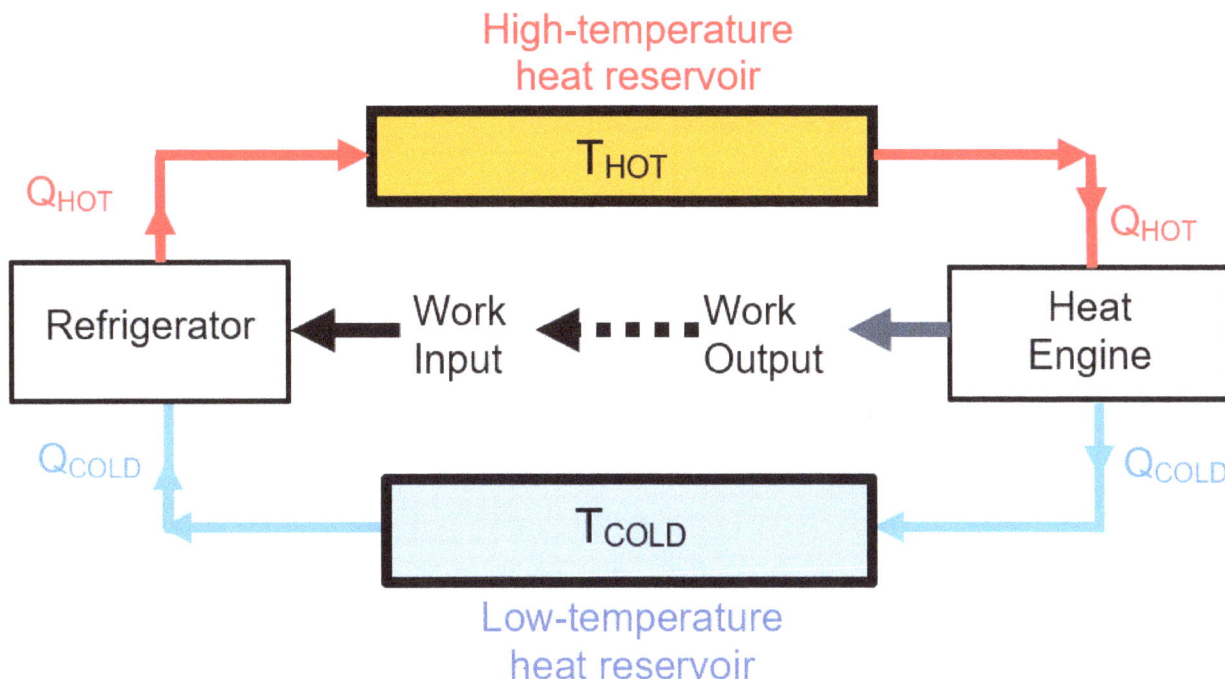

First, let's look at the refrigerator. It removes heat Q_{COLD} from the low-temperature reservoir and transfers it "uphill" to the high-temperature reservoir. As we have seen, if the refrigerator is the most efficient that it could possibly be, the amount of work required to transfer heat Q_{COLD} is:

Equation (1)   Work input to refrigerator $= Q_{COLD}\left[\dfrac{T_{HOT} - T_{COLD}}{T_{COLD}}\right]$

Since energy cannot be gained or lost, all energy inputs to the refrigerator (heat **and** work) must be equal to the heat energy output transferred to the high-temperature reservoir Q_{HOT}. Consequently, the heat transferred to the high-temperature reservoir is:

$$Q_{HOT} = Q_{COLD} + Q_{COLD}\left[\dfrac{T_{HOT} - T_{COLD}}{T_{COLD}}\right]$$

Simplifying and re-arranging, we get:

Equation (2) $Q_{HOT} = Q_{COLD}\left[\dfrac{T_{HOT}}{T_{COLD}}\right]$

Now, let's look at the heat engine, which is as efficient as it could possibly be. The engine takes in heat Q_{HOT}, converts some of Q_{HOT} into work, and dumps remaining heat Q_{COLD} into the low-temperature reservoir. If our hypothetical apparatus is not to violate the Second Law of Thermodynamics, the work output produced by taking heat Q_{HOT} from the heat source and putting it through the heat engine cannot be greater than the work required for the heat pump to push heat Q_{HOT} back into the heat source.

If the engine is as efficient as it could be, its work output equals the work input of the refrigerator (given by Equation 1), using the amount of heat Q_{HOT} (given by Equation 2). So, the fraction of heat energy input Q_{HOT} that the ideal engine can convert into work (it's "efficiency"), is:

Equation (3)

Maximum efficiency of heat engine $= \dfrac{\text{Work output}}{\text{Heat Input } Q_{HOT}}$

$$= \frac{Q_{COLD} \left[\frac{T_{HOT} - T_{COLD}}{T_{COLD}} \right]}{Q_{COLD} \left[\frac{T_{HOT}}{T_{COLD}} \right]}$$

$$= \frac{T_{HOT} - T_{COLD}}{T_{HOT}}$$

And hence,

The maximum work output of a heat engine $= Q_{HOT} \left[\dfrac{T_{HOT} - T_{COLD}}{T_{HOT}} \right]$

Achieving maximum efficiency for a heat engine requires the opposite condition as applies for a heat pump – namely, the temperature difference $T_{HOT} - T_{COLD}$ should be as large as possible. Since T_{COLD} is usually the temperature of the surrounding atmosphere, which we cannot control, *it is critically important that heat engines operate with a heat source at the highest possible temperature*.

Note that the maximum possible Coefficient-of-Performance for a heat pump, and the maximum possible efficiency of heat engine, are ABSOLUTE. It does not matter what technology is used, or how cleverly it is designed. It does not matter if, within the next 200 years, scientists discover entirely new technologies for heat engines. The efficiency of a heat engine will NEVER exceed the limit given in Equation (3) - unless someone figures out a way to violate the Second Law of Thermodynamics (and I doubt this will ever happen). This does not mean that there is no scope for improving the efficiency of heat engines and heat pumps. Even with the best technology that we have now, most heat engines achieve an efficiency which is only about half of the maximum possible efficiency.

With current technology, the efficiency of heat engines is limited mainly by the maximum temperature that can be withstood by the materials used in the engine. Most electricity generating plants operate on a steam cycle. Water is boiled into high pressure steam within a boiler, and the steam is expanded in a steam turbine. The heat source is generally provided by combustion of coal, or by a nuclear reaction. Coal and other fossil fuels typically produce flame temperatures of 1,500°C or more, but steam cycle power plants operate with a steam temperature of about 500°C. At higher temperatures, steel tubes used to contain the high-pressure steam begin to lose strength. To avoid the possibility of the steam tubes bursting (causing a violent explosion and requiring an immediate shut-down and repair of the plant), the steam temperature is generally limited to about 500°C.

In a nuclear power station, the heat source is provided by the nuclear fission reaction occurring in a nuclear reactor. The possibility of a high-pressure steam tube weakening and exploding within a nuclear power plant is even more worrying than in a coal-fired power

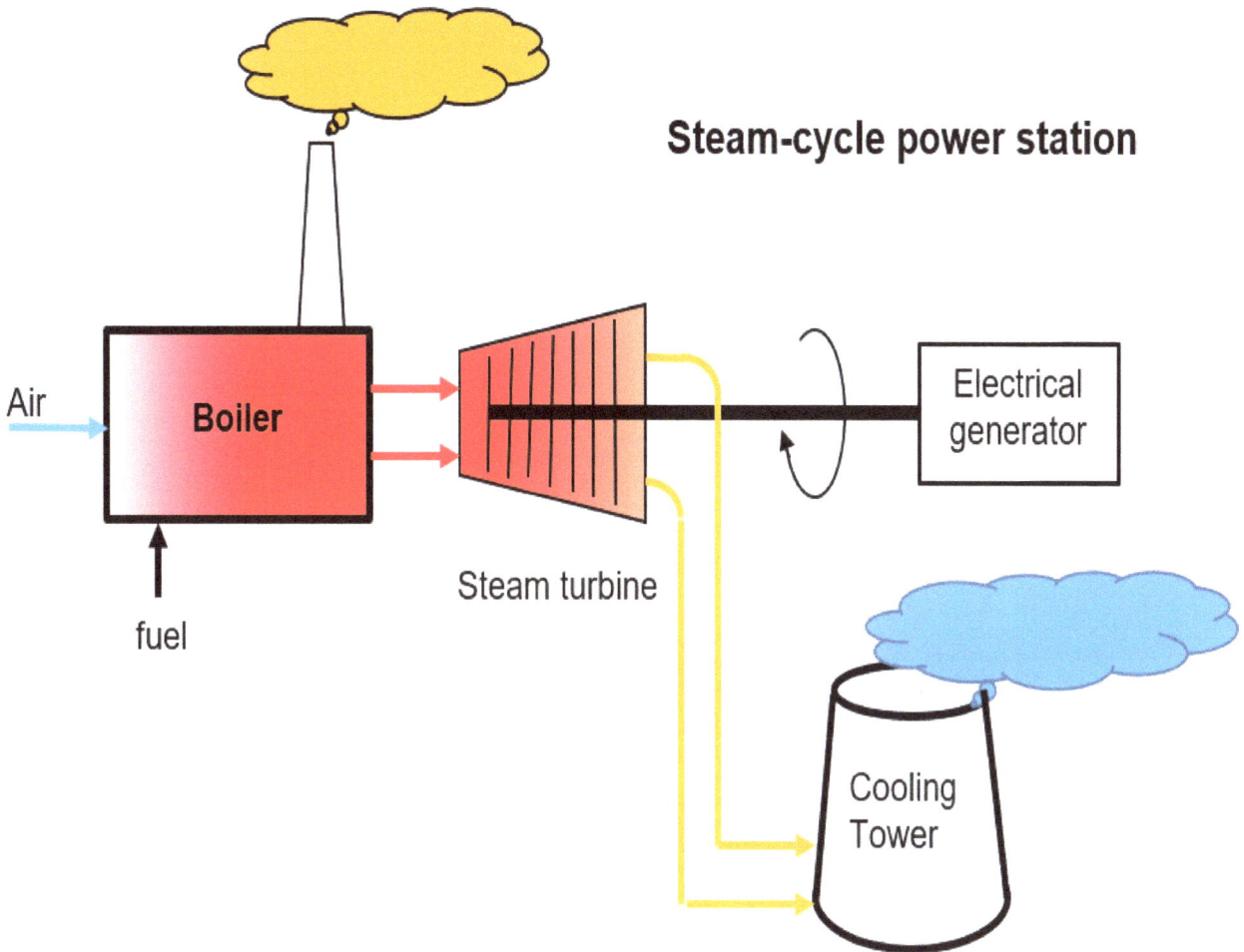

Steam-cycle power station

station, so nuclear power plants are generally designed to operate at slightly lower steam temperature (with a correspondingly slightly lower efficiency).

The maximum possible efficiency of a power station with a steam temperature of 500°C (773° above absolute zero) and air temperature of, say 27°C (300° absolute), is (773-300)/773 = 61%. The best that we could ever hope for would be to convert 61% of the energy released by the burning fuel into useful work output (mechanical power or electricity), The remaining heat (39% of the fuel energy) must be dissipated to the atmosphere.

In actual practice, coal-fired power plants achieve efficiencies of about 35%, slightly more than half the theoretical maximum. Several factors account for additional losses. The boiler cannot recover all of the heat energy released by the burning fuel, so some heat remains in the hot exhaust gases carried up the stack and discarded to the atmosphere. Additional losses occur in the steam turbine and in the generators which convert the mechanical power produced by the steam turbine into electricity. Finally, because of the limited rate at which waste heat can be dissipated to the atmosphere, the steam is condensed back to water at above atmospheric temperature.

To condense steam at the lowest possible temperature, most coal and nuclear power stations use cooling towers, in which the waste heat is absorbed by evaporating water. In Australia, however, coal-fired power stations are often located in arid remote areas, where limited cooling water is available. For a power station with a steam temperature of 500°C, the waste heat that must be dissipated to the environment is about two-thirds (the fraction 0.39/0.61) as much as the power output of the plant. For each kilowatt-hour of electricity produced, two-thirds kWh of heat must be eliminated, requiring one litre of water to be evaporated.

Coal-fired power stations that have been built in remote of areas of Queensland in recent years use "dry cooling", which does not rely on evaporation of water to eliminate waste heat. Rather, these power stations circulate and condense the steam inside metal coils with fins (much like the radiator of your car), using large fans to blow cooling air over the fins. "Dry cooling" is not as efficient as "wet cooling", so the steam is condensed at slightly higher temperature, causing a loss of efficiency or perhaps 1-2%.

Steam-cycle plants normally operate by burning coal or using a nuclear reactor as a heat source. These energy sources are generally much cheaper than natural gas or other fuel sources, but steam-cycle power stations are generally very large (typically, with units of around 1,000 megawatts), and very costly to build. Because of economics and technological limitations, steam-cycle plants are normally operated as "base load" power plants – operating day and night, weekdays and weekends – as much of the time as possible.

Because coal-fired and nuclear steam-cycle plants are costly to build and relatively cheap to operate, utilities want to keep them operating and producing electricity (and earning revenue) as much of the time as possible. Since these plants are expensive to build, their owners pay high financing costs – whether or not the plant is generating electricity. Coal and nuclear fuel are generally cheaper than other fuels, so these plants can operate profitably even if they sell the electricity at discounted off-peak rates. Also, the technology of steam-cycle plants favours base load generation. To bring a massive boiler to operating temperature takes many hours, as rapid heating could cause uneven expansion and crack the steam tubes. Similarly, when the output of a steam-cycle plant is not needed, the boiler must be kept hot or be allowed to cool very slowly. This means that steam-cycle plants cannot be quickly brought on-line, or quickly taken off-line, to meet short surges in power demand.

Consequently, while steam-cycle generating plants are cost-effective in meeting the constant base load of an electricity grid, additional generating capacity is required to meet short-term variations in power demand.

One type of "peak load generating plant" uses gas turbines. Basically, gas turbines are a modified version of jet engines used in commercial aircraft. The main difference is that gas turbines produce mechanical power output in their central rotating shaft, while the power output of aircraft jet engines is used to accelerate the exhaust gases to high velocity (which produces forward thrust). Gas turbine power stations can be started and shut-down quickly, and are relatively cheap to build. However, gas turbine plants are limited to using relatively expensive fuels (such as natural gas or diesel fuel), so they normally operate only during peak demand periods, when a premium rate is paid for electricity sold to the grid.

Burning of natural gas or diesel fuel within a gas turbine produces temperatures in excess of 1,500°C, but here too, engine operation is limited to the temperature range at which component materials retain their strength. In a gas turbine, air is compressed by rapidly-spinning turbine blades, mixed with fuel and burned in combustion chambers, and then the hot exhaust gases are expanded through spinning turbine blades. The very high rotational speed of the turbine imposes high stress on the blades, which must have very high tensile strength to resist being torn apart. Although special metal alloys are used for the turbine blades, which are air-cooled, temperatures within the engine must be kept within the range that the blades can withstand. As a result, much of the heat of the burning fuel is retained in the exhaust gases, which exit the gas turbine at temperatures above 600°C. Consequently, the efficiency of such "open-cycle" gas turbines is generally about the same (or slightly less) than a steam-cycle plant.

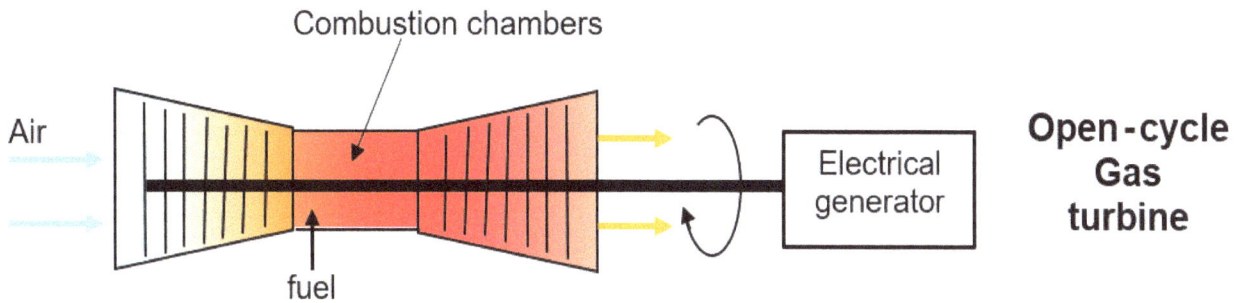

Combustion chambers

Air

fuel

Open-cycle Gas turbine

Electrical generator

However, the "waste heat" in the exhaust gases of a gas turbine is a high-grade heat resource. With temperatures exceeding 600°C, the exhaust gases are hotter than the operating temperature of steam-cycle plants. It makes sense to use the high-grade "waste heat" of an open-cycle gas turbine, and this is exactly what is done in a "combined cycle" gas turbine plant. Here, hot exhaust gases produced by a gas turbine are passed through a boiler, producing high-pressure steam which is then expanded to produce additional power in a steam turbine.

Combined cycle power plants contain a gas turbine and a small steam-cycle power generator, and convert about 60% of the fuel energy into electrical power output. Of course, combined cycle generating plants are more complex and more costly to build than an open-cycle gas turbine plants, but consume much less fuel (and produce much less emissions) to generate the same electrical power output.

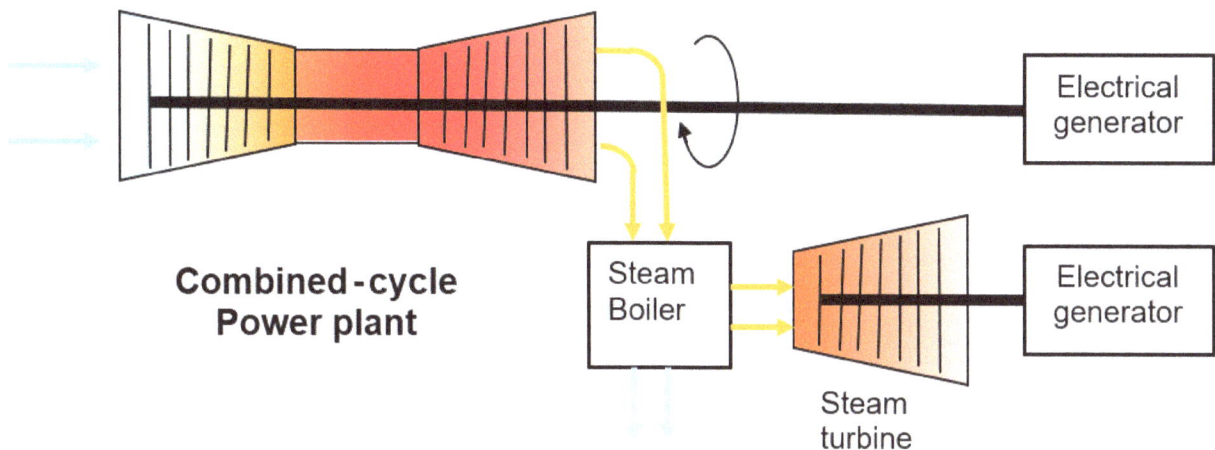

Combined-cycle Power plant

Electrical generator

Steam Boiler

Electrical generator

Steam turbine

In the past, nearly all power plants using natural gas were open-cycle gas turbines, as these were cheap to build (but inefficient and costly to operate). However, in recent years, combined cycle plants have become common.

Low-temperature geothermal power generation

In striving to achieve greater fuel efficiency for electricity generation, engineers design power stations to operate with the highest temperatures allowed by the materials and technology available. However, in some cases, the maximum temperature is determined by the available heat source, which may be low-grade "waste heat". An interesting example is a low-temperature geothermal generating station in the remote town of Birdsville in outback Queensland. The town relies for its water supply on bore water from an artesian aquifer deep underground. The water reaches the surface at nearly 100°C, just below the normal boiling point of water. The hot bore water needs to be cooled before it can be circulated to town residents, so it comprises a free source of unwanted "waste heat". At the same time, electricity for the town was produced by engines burning diesel fuel, a very expensive energy source.

71

A low-temperature Rankine cycle engine was installed – basically a refrigerator operated in reverse, exactly as depicted in the simple diagram at the beginning of this chapter. Hot bore water was passed through a boiler, where liquid propane boiled into vapour. The propane vapour was expanded to produce power, condensed into liquid at air temperature, and then recycled through a pump to the boiler.

In this case, the temperature of the heat source was limited to the 100°C temperature of the bore water, and the heat sink was the local environment at ambient temperature. The air temperature in Birdsville is generally above 30°C in daytime, and often exceeds 40°C in summer. Consequently, the engine had to operate within a narrow temperature range between 40°C (313° above absolute zero) and 100°C (373° absolute). The maximum possible efficiency that could be achieved for such a power station is (373°-313°)/373° = 16%.

In fact, the actual efficiency would be significantly less than 16%. The power station contained a single heat engine to minimise complexity and cost, so its boiler extracts heat from the hot bore water at a single temperature. What would be the best temperature to operate the boiler? Let's consider two options:

- We could extract heat from the hot water at, say, 98° – just below the temperature of the bore water. This would allow the highest possible efficiency of the heat engine, but we would only extract a tiny fraction of the heat energy in the water as it cooled from 100° to 98°.

- Alternatively, we could operate the boiler at, say, 42°C – just above ambient air temperature. This would allow the boiler to extract nearly all of the heat energy in the hot water, but the efficiency of the engine would be extremely low.

It should be evident that maximum power would be produced by extracting heat at a temperature roughly halfway between 100°C (the available T_{HOT}) and 40°C (T_{COLD}). For readers who completed first-year university calculus, it is a nice exercise[See Note 1] to derive that the optimal temperature that allows maximum power production is 68°C. At this boiler temperature, a maximum of 6.7% of the heat content of the bore water can be converted into electrical power output.

Each kilolitre of water (1,000 litres) releases 250 megajoules of heat as it cools from 100°C to 40°C. As much as 6.7% of this might be converted into work output. Thus, the heat contained in a thousand litres of hot bore water could potentially yield up to 4.6 kilowatt-hours of electricity.

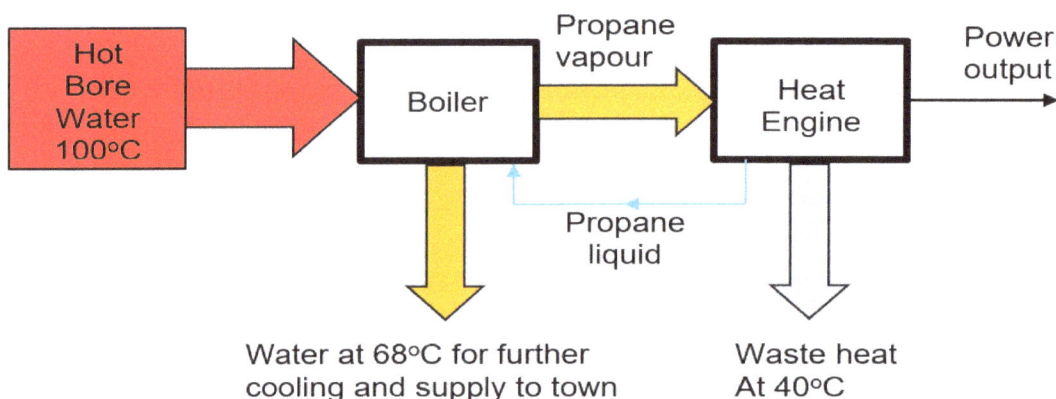

Configuration for extracting heat from hot bore water using a single heat engine. The width of the arrows indicates the relative energy content in the stream.

The actual efficiency and power output of the power plant would be even less than this. However, since the heat source is literally available at zero cost (as the water needs to be cooled anyway), and the alternative diesel generation is very expensive, low-temperature geothermal power generation proved cost-effective and operated for many years in Birdsville.

Solar power generation

Throughout the 4.5 billion year existence of the Earth, sunlight has been the major energy source for the planet and its lifeforms. Solar energy was used by ancient plants and algae living in shallow seas to produce carbohydrates, proteins and oils. Some of this biomass was converted into coal, oil and gas when deposits accumulated and were buried in sediments as the seafloor subsided over millions of years.

Sunlight creates temperature differences across the Earth's surface, and the atmosphere acts as a heat engine to redistribute heat from equatorial regions towards the poles. This gives rise to winds, waves, rains and storms.

In recent years, solar energy has been tapped directly as an energy source for humanity. Two types of technologies are used to generate electricity from sunlight:

- Solar thermal power generators concentrate sunlight onto a collector, with the collected heat converted to mechanical power (and then, electricity) in a heat engine.

- Photovoltaic power generators convert solar radiation directly into electrical energy.

We can think of the entire solar system as a huge heat engine with the sun as a high-temperature heat reservoir. The surface of the sun has a temperature of about 6,000° (absolute) and each square metre emits about 70 million watts of visible, ultraviolet and infrared radiation. As the radiation moves outwards and spreads across the solar system, its intensity reduces with the inverse square of the distance. As it travels from the sun's surface (at a radius of 700,000 kilometres) to the orbit of the Earth (at a radius of 150 million kilometres), the light spreads across an area some 40,000 times larger. By the time the radiation reaches Earth, its intensity has fallen to about 1,300 watts/m^2. Some is scattered by the Earth's atmosphere, and most ultraviolet radiation is absorbed, so that sunlight striking perpendicular to the Earth's surface has an intensity of about 1,000 watts/m^2.

It is possible, using mirrors or lenses, to concentrate sunlight to increase its intensity. High temperatures are produced when concentrated solar radiation is absorbed by a collector plate. In theory, it would be possible to produce temperatures as high as 6,000° – as hot as the temperature at the surface of the sun. It is not possible to produce even higher temperatures by cleverly focussing and concentrating solar radiation, as this would transfer heat from the surface of the sun to an even hotter collector on the Earth – and this would violate the Second Law of Thermodynamics.

Mind you, with current technology, it would not be possible to build a solar collector that could withstand 6,000°. All known materials vaporise or decompose at temperatures well below this. Nonetheless, it is instructive to consider what would happen if new materials could be developed in the future that could withstand such temperatures.

As we focussed highly concentrated solar radiation on our hypothetical absorber, its temperature would rise and it would emit "blackbody radiation" from its surface - just as an object heated in a furnace begins to glow red hot. The intensity of emitted radiation rises rapidly with increasing temperature. Each 1% increase in absolute temperature increases the intensity of emitted radiation by 4%. As the temperature of the absorber approaches 6,000°, the intensity of emitted radiation becomes nearly as intense as the highly concentrated sunlight. If the "absorber" reached 6,000°, it would emit radiation (and lose energy) at the same rate as it absorbed concentrated sunlight. At even higher temperature, the "absorber" would emit more radiation than it absorbed, and its temperature would fall.

Let's consider the maximum possible efficiency of a solar thermal power station, which is basically a heat engine using a solar collector as a heat source. The maximum temperature of the high-temperature heat reservoir is 6,000° (the temperature at the surface of the sun), and the low-temperature heat reservoir is the surrounding environment on Earth (at a temperature of around 300° absolute). Consequently, it would ultimately be possible for **absorbed solar radiation** to be converted into useful work with an efficiency of (6,000-300)/6,000 = 95%. But this does not mean that 95% of **incident solar radiation** can be converted into work. A solar thermal power station faces the same type of dilemma as we encountered with a low-temperature geothermal power station. The question to consider is this: What is the optimal temperature at which solar radiation should be absorbed?

To illustrate the problem, let's again consider two hypothetical options:

- We could operate the solar collector at low temperature – just above ambient temperature. Then, radiation emitted by the collector would be negligible, and nearly all the incident solar energy would be available to operate the heat engine. But, at this low temperature, the collected heat could only be converted at low efficiency into useful work.

- On the other hand, we could operate the solar collector at just below the maximum possible temperature of 6,000°. In principle, an ideal heat engine could convert this heat into work with 95% efficiency. But, at this temperature, the collector would be emitting radiation at nearly the same rate as the concentrated solar radiation was being absorbed. The net heat provided to the heat source would be only a small fraction of the incident solar energy.

Once again, It should be evident that maximum power would be produced by operating the solar collector at some intermediate temperature, well below the theoretical maximum of 6,000° and well above ambient temperature. As it turns out that a maximum of 86% of incident solar energy could (in theory) be converted into electricity, and this would be achieved at a collector temperature of around 2,500° (absolute).

In fact, with current technology, the actual efficiency of solar thermal power stations is about one-third of the maximum possible efficiency, so there is still considerable scope for technological improvement.

Getting around efficiency limits

The fundamental limit on the efficiency of heat pumps and heat engines is a consequence of the Second Law of Thermodynamics. It is most unlikely that any technology will **ever** be developed that can "beat" this efficiency limit. However there are ways to circumvent the efficiency limit that applies whenever heat is converted into work.

Heat engines operate by continuously absorbing heat from a high-temperature reservoir, converting some of the heat to work, and rejecting the remaining het to a low-temperature reservoir. Nearly all engines work by burning fuel (coal, oil, gas or biomass), converting the chemical energy stored in the fuel into heat. But it is also possible to convert the chemical energy in a fuel directly into electricity - without first converting the chemical energy to heat. This occurs in fuel cells, which are not subject to the efficiency limit that applies to heat engines.

Fuel cells have been used to power spacecraft, and in other niche applications. For many years, fuel cells were expected to gain wide acceptance as a power source for cars and vehicles, but fuel cell developers have not yet been able to overcome technological shortcomings. Firstly, current fuel cells are essentially limited to using hydrogen or natural gas as an energy source. Furthermore, the most efficient fuel cells convert only about half of the fuel energy into electricity. But these limitations are due to shortcomings of existing fuel cell technology – not due to any fundamental limitation, and might be overcome as improved fuel cells are developed.

Notes

1. We want to produce maximum power from bore water containing heat energy Q_{HOT}. Let's say that the boiler extracts heat from the bore water at some particular temperature T, which is intermediate between T_{HOT} and T_{COLD}. The amount of heat recovered from the bore water is then $Q_{HOT}\left[\dfrac{T_{HOT}-T}{T_{HOT}-T_{COLD}}\right]$. Of this heat energy, the fraction that can be converted into useful work is limited to $\left[\dfrac{T-T_{COLD}}{T}\right]$, so the maximum work output is $Q_{HOT}\left[\dfrac{T_{HOT}-T}{T_{HOT}-T_{COLD}}\right]\left[\dfrac{T-T_{COLD}}{T}\right]$.

    We can find the value of temperature T that yields the highest work output by taking the first derivative of the power output and setting this equal to zero. Solving for the optimal value of T gives $\sqrt{T_{HOT}\,T_{COLD}}$. If the temperature difference $T_{HOT} - T_{COLD}$ is small in relation to the absolute temperature T_{HOT}, then it turns out that the optimal temperature is halfway between T_{COLD} and T_{HOT}. Substituting this optimal temperature for the conditions $T_{HOT} = 100°C$ and $T_{COLD} = 40°C$ gives a maximum power output equal to 6.7% of the heat energy in the stream of hot water.

10. Fresh water and the challenge for humanity

Perhaps the greatest challenge facing humanity in the 21st century is providing food for a growing world population, in the face of a range of major threats – depletion of fish stocks, the effects of climate change on agriculture, depletion of ground water resources, salinization of soils, desertification, and loss of prime agricultural land to development. Increasing crop losses may occur as insect pests develop resistance to existing insecticides, and weed species are likely to develop resistance to herbicides that are currently used to control them.

While the human population has stabilised or is even shrinking in some countries, the population continues to grow overall, especially in Africa and parts of Asia with high birth rates. Depending upon which population projections you believe, the population of the Earth in the year 2100 will likely be 30-60% more than at present. At the same time, several billion people are expected to rise from poverty into the middle class, and they will expect a better and more diverse diet (probably including more meat), and this will put even greater pressure on agricultural systems.

Probably the most important factor which will limit food production is the availability of fresh water. Enormous amounts of water are required to grow crops. Globally, three-quarters of all water consumed by humanity is consumed in agriculture. The amount of water "embodied" in growing the food that you consume is probably more than twice as much as you consume in your home (for drinking, cooking, dish and clothes washing, showering, washing, toilet flushing and gardening).

All terrestrial plants lose water by evaporation in their leaves when they undertake photosynthesis. Whenever leaves absorb carbon dioxide from the air, as they must do to produce carbohydrate and grow, water is lost by evaporation through the membranes in plant cells. This causes water to be transported from the roots of the plant, carrying nutrients from the soil and preventing the leaves from over-heating. Depending upon the type of plant, about 300-600 litres of water are lost by "transpiration" of water vapour from their leaves for each kilogram of (dry) carbohydrate produced. The amount of water used to grow crops is often much greater than this. Only a fraction of the carbohydrate produced by the plant ends up in the edible fruit or vegetable that gets eaten, with the remaining carbohydrate going into the stems, leaves and other inedible parts of the plant. Often, substantial water losses occur through run-off and evaporation before the water reaches the roots of the plant.

To maintain and expand agricultural output, farms in many countries increasingly rely on irrigation, often using ground water extracted from wells. Ground water provides about half of the drinking water in the United States, and more than 200 trillion litres/day (200 cubic kilometres) for agriculture [Ref 1]. However, in the US and many other countries, ground water is being consumed much faster than it is being replenished, with the level of ground water falling to increasing depths. Increasing amounts of pumping energy are used to extract irrigation water from greater and greater depths. One study [Ref 2] estimated that 40% of the water withdrawn from aquifers is not replenished, and that the global rate of ground water depletion is about 280 trillion litres/year. Severely affected areas include Northeast Pakistan, Northwest India, Northeast China, the central U.S, California, Iran, Yemen, and Southeast Spain. Water that is withdrawn from underground aquifers eventually ends up in the oceans (either after evaporating and condensing as rain, or by run-off into streams), and is responsible for about one-fourth of the current rate of sea-level rise (3.1 mm per year).

Pumping out ground water also changes the course of underground streams, sometimes causing them to flow through sediments containing arsenic or other toxins. At least 140 million people in Asia are already drinking arsenic-contaminated water. At excessive levels, arsenic causes brain damage, heart disease and cancer. Arsenic-contaminated ground water has been found in at least 30 countries.[Ref 3]

Irrigation water is also withdrawn from rivers and dams, but many river systems are now fully exploited, with little scope to draw additional water to expand agricultural production. Some rivers are over-exploited, leaving little or no residual flow to maintain the environmental health of the river (to flush out sediment or salinity, or prevent seawater from moving upstream from the mouth of the river). Allocation of water rights between farmers, commercial interests and town water supply authorities may cause disputes and acrimony, as has occurred in the Murray-Darling River system.

Here is a table of irrigated crops grown in the Murray-Darling Basin [Ref 4].

	Typical water use per hectare	Crop area hectares	Production Tonnes per hectare	Calculated water use kilolitres per kilogram
Almonds	14 ML/ha	40,000	2 tonnes/ha	7,000 litres/kilogram
Rice	12 ML/ha	80,000	8 tonnes/ha	1,500 litres/kilogram
Cotton	8 ML/ha	394,000	2.5 tonnes/ha	3,200 litres/kilogram
Grapes	9 ML/ha	8,000	15 tonnes/ha	600 litres/kilogram
Oranges	6 ML/ha	20,000	24 tonnes/ha	250 litres/kilogram

Consider that, when you buy a kilogram of rice grown in the Murray-Darling Basin, about 1,500 litres (1.5 tonnes) of water was used to grow the product. If rice farmers in the Murray-Darling Basin paid the same water charges as do residents of Brisbane ($3.57/kilolitre), this would add $5 to the production cost for each kilogram of rice. With a retail mark-up of, say, 200%, this would increase the supermarket price for a kilogram of rice by about six times!

When you are munching on almonds, consider that about 8 litres of water were used in growing *each* almond. About 500 litres of water (0.5 tonnes) is used to grow the cotton in one polo shirt.

Obtaining sufficient water to grow crops is not a new problem. Since the onset of agriculture about ten thousand years ago (and probably before then), human societies faced periodic episodes of drought. Severe or prolonged droughts led to starvation, social unrest, war and population collapse. We now know that great civilisations, containing cities and empires with hundreds of thousands or millions of people, arose in ancient times, prospered for centuries and then collapsed or declined. These societies had complex and sophisticated systems of specialised labour and social organisation. Many constructed elaborate temples, pyramids, monuments or other complex structures requiring thousands of workers, administrators and a religious/political hierarchy. But ultimately, all societies are based on a reliable production of sufficient food. For many of the great civilisations of the ancient world, prolonged or severe drought is implicated as a major cause of their collapse or decline.

Could lack of fresh water threaten human civilisation in the 21st century? For one thing, virtually the entire human population is now connected and integrated through global trade. So, crop failures due to localised drought (or flood) in one area could be offset by food imports from other unaffected areas (providing that local people have the financial resources or reserves to purchase food imports). On the other hand, the close interconnectedness and interdependence of the modern world leaves us more vulnerable to global climate disruptions.

Now, for the first time in history, modern technology allows fresh water to be extracted from seawater. Desalination of seawater can supplement, or even replace, natural rainfall for supply of fresh water. Any country or region that has access to the coast, and to the world's oceans (containing an estimated 97% of all water on Earth), has the capacity to produce virtually unlimited amounts of fresh water. However, there is a "cost" to large-scale production of fresh water from the sea – a cost in financial terms, and in the energy required to extract fresh water from seawater. As we shall soon see, there is a fundamental minimum energy requirement to extract fresh water from the oceans, and that energy is directly related to the salt concentration of seawater. The scope for human society to sustain agricultural production, and to expand food production to meet the needs of a growing population, will depend on this fundamental technological limit.

Distillation

First, let's consider how nature provides the rain (and snow) that is the source of fresh water in lakes, streams, reservoirs, glaciers and ground water. Nature's supply of fresh water is provided by evaporation of seawater and its subsequent condensation as rain.

Water exists as a liquid at normal temperature and pressure because its constituent molecules are attracted to each other by intermolecular forces. These attractive forces hold the water molecules close to their neighbours in the liquid state. Very few water molecules have sufficient energy to overcome these attractive forces and break free – but some do! Water molecules are continually exchanging energy by colliding with one another, and this causes some molecules to have more energy than average (and other molecules have less energy). At room temperature, about one molecule in ten million has sufficient energy to break free of these attractive forces, and escape into the vapour.

Imagine that a small amount of water is placed into an empty container. Initially, we pump out all gases in the container, bringing the gas pressure to zero. However, as soon as we turn off the pump, the pressure begins to rise as water molecules escape from the liquid. The pressure of this water vapour increases, but soon stabilises at a constant value. At this "equilibrium vapour pressure", water molecules are escaping from the liquid at the same rate as water molecules hit the water surface and return to the liquid. For water, the equilibrium vapour pressure is 3 kilopascals (3% of atmospheric pressure) at 25°C. The equilibrium vapour pressure is the same, regardless of how much or how little water is present in the container. In fact, *for pure water (or any pure liquid), the equilibrium vapour pressure depends only on the temperature*. However, even a small change in temperature causes a huge change in the equilibrium vapour pressure. Roughly speaking, a 10°C rise in temperature causes the equilibrium vapour pressure to double.

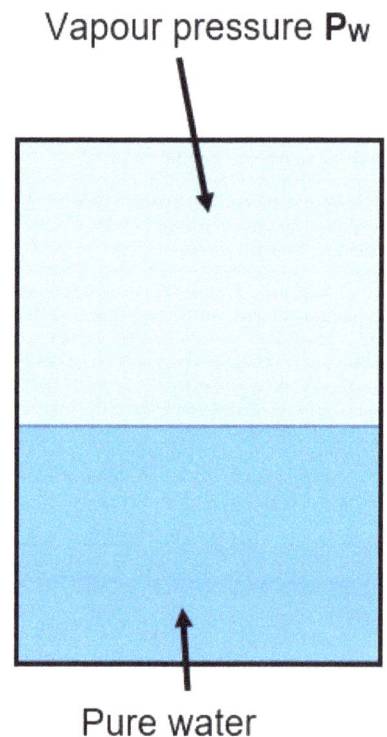

Vapour pressure P_W

Pure water

Water is a volatile liquid – that is, it has a significant equilibrium vapour pressure at "normal" temperatures. But the salts contained in seawater (mostly sodium chloride) are not volatile: only water molecules escape from the liquid during evaporation. Salt remains in the liquid solution. This process is simulated in a laboratory by distillation. We need only to add heat to a vessel containing salt water (increasing the vapour pressure of the water), and to allow the water vapour to pass into a vessel at lower temperature, where the vapour condenses to water.

For each mole of water that we evaporate, we need to provide its Molar Heat of Vaporisation E_{vap}. Consequently, simple distillation requires huge amounts of energy. Each mole of water that vaporises absorbs about 40,000 Joules of heat. The same amount of heat is released when the vapour condenses into water, but with simple distillation, heat released as the vapour condenses escapes into the environment.

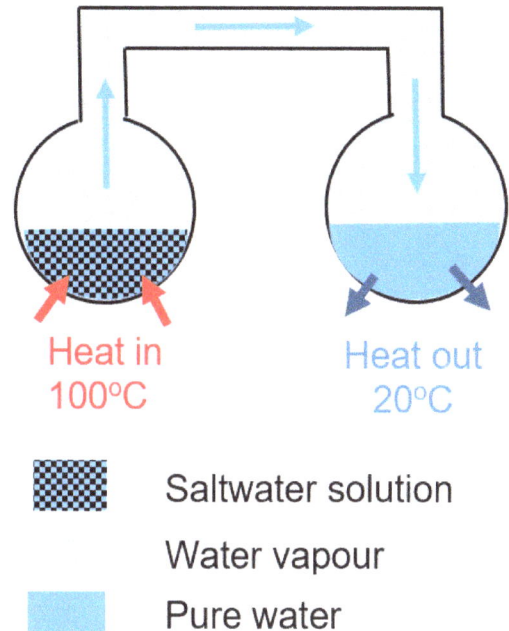

Heat in
100°C

Heat out
20°C

Saltwater solution

Water vapour

Pure water

A litre of water contains 55.5 moles so, for each litre of water that vaporises absorbs 2.2 million Joules (that is, 2.2 Megajoules) of heat. This amount of heat could be provided by, say, burning 0.1 kilograms of coal. That might not sound like a big deal, but if we are looking to grow food crops by distilling seawater, *we would need to burn 150 kilograms of coal to produce enough fresh water to grow one kilogram of rice !* This would be completely impractical, economically unviable and would have disastrous environmental impacts.

In the natural world, distillation of seawater provides all the fresh water that falls as rain or snow. But the Earth has millions of square kilometres of oceans and land surface to act as an enormous collector of solar energy. Consequently, a huge amount of solar heat is available to drive evaporation of seawater.

We can calculate the total amount of solar energy absorbed by the Earth's surface and determine the amount of water that can be evaporated. When we compare this with the total rainfall across the Earth's surface, the two answers are nearly the same!!! [note 5] This suggests that water vapour carried by air currents is a major mechanism for distributing heat around the planet (with evaporation occurring near the equator, and water vapour condensing into cloud droplets as moisture-laden air currents move towards the poles or to higher altitude).

Although simple evaporation of seawater and its condensation into fresh water is the source of all the lakes, streams, rivers, glaciers and groundwater, it is an inefficient process for human society to supplement natural rainfall. Using fossil fuels to provide the 2.2 megajoules of heat needed to evaporate each litre of water would be extremely energy-intensive and costly, and completely impractical as a source of water for agriculture.

In many countries, total use of fresh water (for agriculture, industry and households) is about 3,000 litres per person per day. To provide this amount of water by simple distillation would require 250 kilograms of coal or 200 litres of diesel fuel for each person every day. This is about ten times more energy than is used by an average person in western society!

Multi-stage evaporation

By carrying out distillation as a more complex multi-stage process, it is possible to utilise heat released when the vapour condenses. Heat released by condensation in one "stage" can be recycled to evaporate water in the next stage. By using two, three or more evaporation stages, the energy required to produce each litre of water can be reduced by a factor of two, three or more. At each stage of the evaporation process, the temperature and pressure are reduced so that heat can flow "downhill" from one stage to the next. "Multi-stage evaporation" is widely used in Australia's sugar cane industry to concentrate the sugary solution of "cane juice" produced when sugar cane stalks are crushed. By using 4, 5 or 6 stages, the energy requirement can be reduced (relative to the 2.2 MJ/litre for simple distillation) by a factor of 4, 5 or 6 times respectively. While it would be theoretically possible to have as many as 50 evaporation stages, practical considerations limit the number of stages. Thus, while multi-stage evaporation requires about one-fifth as much heat as distillation, it is still energy-intensive.

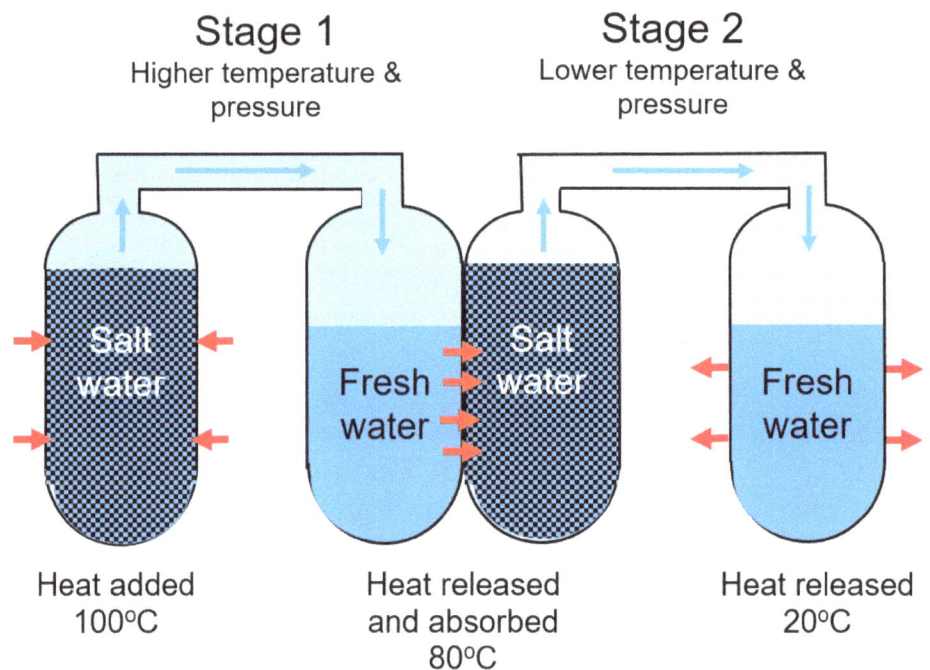

Stage 1
Higher temperature & pressure

Stage 2
Lower temperature & pressure

Heat added
100°C

Heat released and absorbed
80°C

Heat released
20°C

The ultimate distillation process is "mechanical vapour recompression", the "Rolls Royce" of distillation, which vastly reduces the energy required to extract pure water from seawater. An ideal "mechanical vapour recompression" system (that is, an idealised system with no friction, no losses and perfect heat transfer) would be the most energy-efficient distillation system that *could ever* be invented. But even here, there is a lower limit to the energy required to produce fresh water.

Water purification with mechanical vapour recompression

Imagine that we conduct another "thought experiment". This time, we take a sample of saltwater from the ocean and put it in a container. Once again, we pump away any air that is originally present, shut off the pump, and measure the pressure of water vapour above the solution. Let's say that we put this container next to the one containing pure water, and ensure that both containers are at exactly the same temperature.

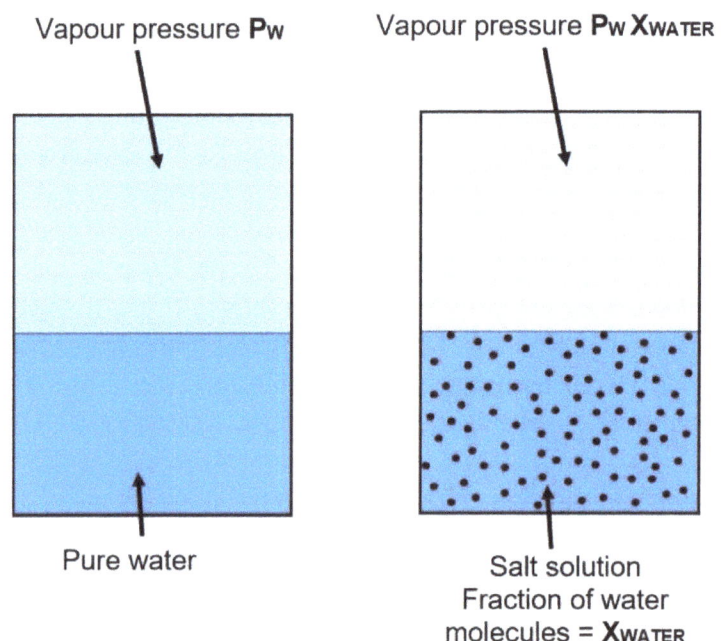

Vapour pressure P_W

Vapour pressure $P_W X_{WATER}$

Pure water

Salt solution
Fraction of water molecules = X_{WATER}

In the container containing pure water, we measure the equilibrium vapour pressure, which we'll give the symbol **Pw**. In the container with salt water, the water vapour pressure is slightly less. Why is this?

Seawater is a solution containing about 96.5% water, and 3.5% salt by weight. Salt consists predominantly of sodium chloride. The sodium chloride does not consist of neutral molecules, but is "dissociated" into sodium ions (Na^+) and chloride ions (Cl^-).

Imagine that we could shrink ourselves to molecular size and survey molecules of water and ions of sodium and chloride as they drift past. Just over 1% of the particles that we count would be sodium ions, just over 1% would be chloride ions, and about 98% would be water molecules. Sodium and chloride ions comprise about 2% of all the particles in solution, so we say that the "mole fraction" of dissolved salt (X_{SALT}) is 0.02. The "mole fraction" of water (X_{WATER}) is 0.98. ***Since 98% of the particles at the surface are water molecules, the vapour pressure of seawater is 98% that of pure water at the same temperature***.

Imagine that we connect the two containers together so that water vapour can freely diffuse and be exchanged between the two solutions.

Due to its salt content, sea water has lower vapour pressure than pure water, creating a difference in pressure between the vapour in the two flasks. Fresh water evaporates to maintain its equilibrium vapour pressure. At the same time, water vapour condenses in the salt solution, which has a lower equilibrium vapour pressure. The net result is that water continually evaporates or boils from the pure water and condenses into the saltwater. This causes heat to be absorbed by the fresh water as it evaporates, and heat to be released when the vapour condenses into the salt water.

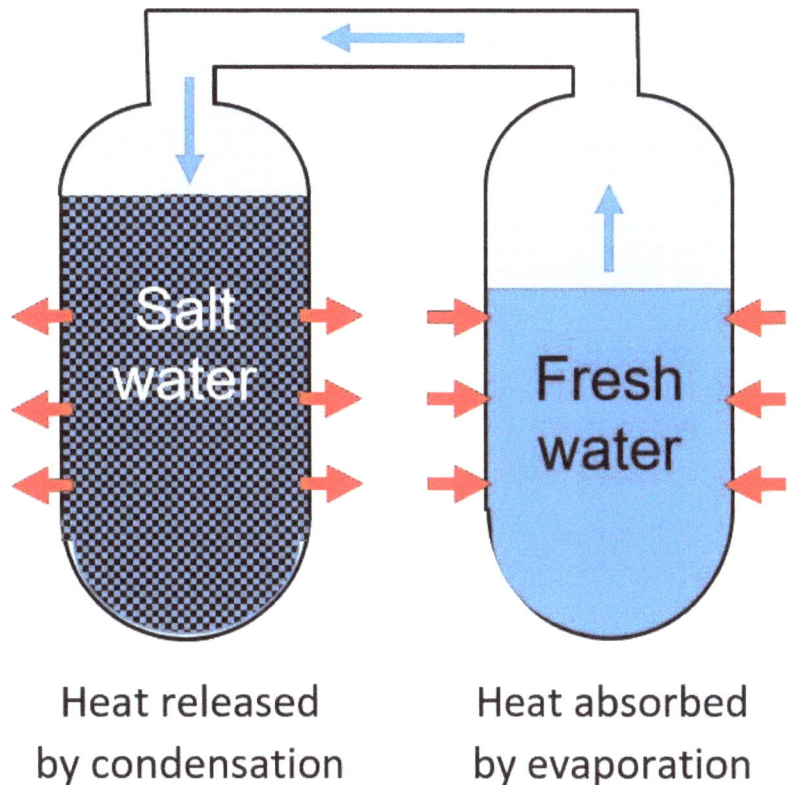

Heat released by condensation   Heat absorbed by evaporation

The net result is that water molecules transfer from the pure water into the saltwater. But this is the ***opposite*** of what we want to happen! If we are trying to desalinate seawater, we need to reverse the direction of the water vapour. We must force water to evaporate from the salt solution and to condense in the flask of pure water. Using a pump, we can reduce the pressure of water vapour above the salt solution, causing it to boil; compress the water vapour to higher pressure; and then push the water vapour into the container of fresh water.

But simply forcing vapour to evaporate from the salt solution and condense in the fresh water creates a problem! Imagine a pump sucking vapour from the salt water flask (causing it to boil), and pushing the vapour into the fresh water flask (causing the vapour to condense).

Evaporation causes the fresh water solution to get colder and colder, while condensation makes the salt solution hotter and hotter. The pump has to work harder and harder to push against the increasing difference in vapour pressure.

However, we can avoid this problem by allowing heat to transfer freely between the two solutions. This maintains both solutions at the same temperature. Then, the heat required to boil or evaporate the saltwater is provided by heat released when vapour condenses into fresh water.

Pump

Water vapour
at pressure P_w

Water vapour at
pressure $P_w X_{WATER}$

Allowing heat to transfer between the two solutions means that we don't need to provide any net heat input. But we must provide mechanical energy to operate the pump, since the pressure of water vapour above the fresh water is more than the pressure above the salt solution.

At whatever temperature we choose to operate this system, pure water has some particular equilibrium vapour pressure, which we'll call P_w. The vapour pressure of the seawater is reduced in proportion to the mole fraction of salt in the solution, X_{SALT} (which is about 0.02 for seawater).

In the container of seawater, the pressure of water vapour is $P_w X_{WATER}$. In the container of fresh water, the pressure of water vapour is P_w. *To transfer water vapour from the seawater into the fresh water, the pump must compress the vapour from an initial pressure of $P_w X_{WATER}$ to a final pressure P_w.* The pump serves the same purpose as the pump in a refrigerator. In mechanical vapour recompression, the pump works against a pressure difference arising from different concentrations of water in the two solutions. In a refrigerator, the pressure difference results from the difference in temperature of the liquid refrigerant.

We have seen that the work required to compress **n** moles of any gas from initial pressure $P_{Initial}$ to final pressure P_{Final} is:

$$\text{Work to compress a gas} = n \, R \, T \, \ln\left[\frac{P_{Final}}{P_{Initial}}\right]$$

Where **n** is the number of moles of gas or vapour
R is the Universal Gas Constant, 8.3 Joules/mole-degree
T is the temperature of the gas, degrees Kelvin

Keep in mind that this is the **absolute minimum energy** that would be required to compress a gas if we had a perfect, frictionless pump, and the gas is compressed very slowly (at constant temperature).

Since the final pressure (in the container of fresh water) is P_w, and the initial pressure (in the container of seawater) is $P_w X_{WATER}$, the minimum work required to compress and transfer n moles of water vapour is:

$$\text{Work to compress water vapour} = n\,R\,T\,\ln\left[\frac{P_w}{P_w\,X_{WATER}}\right]$$

Cancelling out Pw and re-arranging gives:

Eq (1)   Work to compress water vapour $= -\,n\,R\,T\,\ln X_{WATER}$

Note that the fraction of water molecules and the fraction of salt ions must add to one, so:

$$X_{WATER} + X_{SALT} = 1.0 \qquad\qquad \text{So, } X_{WATER} = 1.0 - X_{SALT}$$

And since, the mole fraction of salt is much less than one, the following mathematical simplification applies:

$$\ln(1 - X) = -X \qquad\qquad \text{Where X is much less than one}$$

This simplification is valid (with an accuracy of 90%) if X is less than 0.2. For seawater, in which $X_{SALT} = 0.02$, the substitution of $-X_{SALT}$ for $\ln(1-X_{SALT})$ is 99% accurate.

Inserting this simplification into Equation (1) gives the minimum work required to transfer each mole of water vapour from seawater to fresh water.

Eq (2)   Minimum work required$= R\,T\,X_{SALT}$

This is a very simple result. We know the value of **R**; we know the mole fraction of salt in seawater X_{SALT}. Ideally, we would like the process to occur at room temperature **T**, although to make the process occur more quickly, it is often carried out at higher temperature. This means that the minimum work required to produce one mole of fresh water by mechanical vapour recompression of seawater is 54 Joules, or about **3,000 Joules for each litre of fresh water produced**. This is about **600 times less** than the energy required for simple distillation! Note, however, that this is a theoretical ideal minimum energy consumption that we can only hope to approach with the best possible design, materials and operating conditions. Even this absolute minimum energy required to desalinate seawater under idealised, perfect conditions is still quite significant. It is about the same energy as is required to pump water to a height of 300 metres (about the height of a 100 storey building).

The actual situation is considerably worse than this. As soon as we begin to extract water from a salt solution, its salt concentration (X_{SALT}) increases. The energy required to extract each additional litre of water increases as the remaining solution becomes more concentrated. Once half of the water has been extracted, the remaining solution has twice its initial salt concentration, and we must expend twice as much energy (6,000 Joules) to extract an additional litre (with an average energy expenditure of 4,200 Joules/litre). And then, we must dispose of the remaining concentrated salt solution, usually by pumping it far out at sea.

With the most efficient desalination technology now available, about 30,000 Joules of fuel energy are needed to produce each litre of fresh water. This is about ten times as much as the absolute theoretical minimum.

Two short videos, showing the concept of Mechanical Vapour Recompression, can be viewed at:

https://www.youtube.com/watch?v=L7jvJqT_Ayc

https://www.youtube.com/watch?v=CsBLr32cn8o

Minimum energy to extract water from seawater versus % water extracted

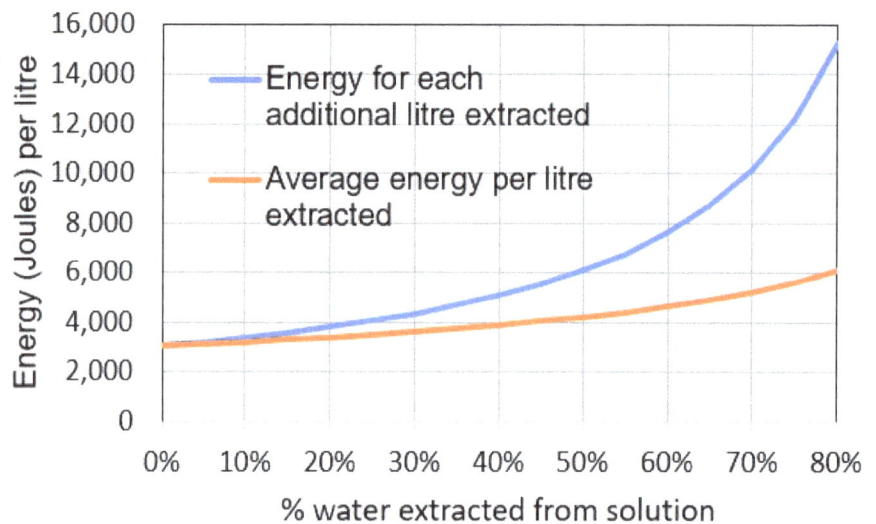

Reverse osmosis

Until the mid-20th century, desalination of seawater was limited to a few niche applications, such as providing fresh water to navy ships, and mechanical vapour recompression was the pre-eminent desalination technology of that era. However, development of synthetic membrane technology in the 1960s opened the possibility for a new desalination process called "reverse osmosis", which is now the dominant technology used for desalinating seawater in Australia and around the world. It relies on a completely different process than distillation or mechanical vapour recompression. It does not involve transitions between liquid and vapour, but occurs entirely within the liquid state. But, as we shall see, the minimum energy required to desalinate water with reverse osmosis is exactly the same as that for mechanical vapour recompression.

The process of "osmosis" has been understood for a long time. It is based on the ability of many membranes, both natural and man-made, to selectively allow particular molecules (notably water) to pass through the membrane, while blocking the passage of other molecules or salts. These membranes can be extremely selective. Microscopic pores in the membrane have the correct size, shape and surface properties to allow water molecules to freely pass to the other side of the membrane, while completely blocking the passage of sodium and chloride ions (or other salts).

We can use such a "semi-permeable membrane" like a "molecular sieve" to filter out salts, bacteria, viruses and other unwanted purities, allowing only pure water to pass.

Imagine that we place seawater and pure water on opposite sides of a semi-permeable membrane. On the fresh water side, any molecule of water that happens to wander into a pore diffuses through the opening and emerges on the opposite side of the membrane.

Osmosis

Semi-permeable membrane

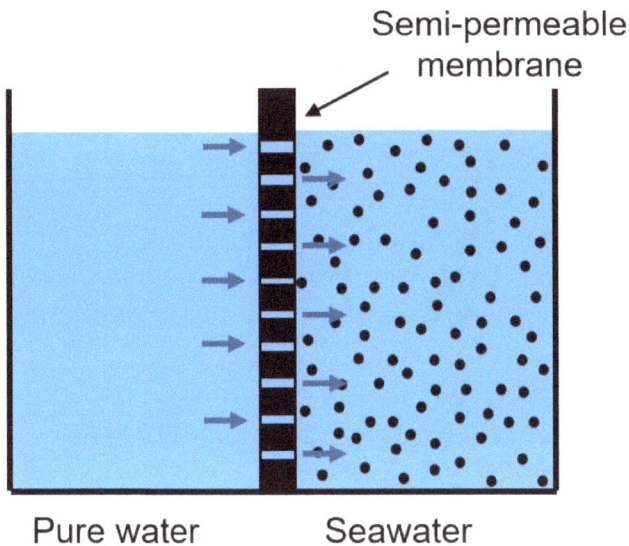

Pure water   Seawater

The same applies on the seawater side of the membrane. There too, any molecule of water that happens to wander into a pore passes through the membrane and emerges on the other side. *But* don't forget that water molecules comprise only about 98% of the particles in seawater. Sodium ions and chloride ions, comprising 2% of particles in seawater, are blocked from passing through the membrane pores. Consequently, more water molecules pass from the fresh water into the seawater than in the opposite direction, so there is a net migration of water from the seawater into the fresh water. This process, called "osmosis", is the *opposite* of what we want to happen if we are trying to extract pure water from seawater!

However, we can stop – and then reverse – migration of water molecules into the seawater by applying pressure across the membrane. When a pressure difference is applied across the membrane, the migration of water molecules into the seawater slows until, at a particular pressure (the "osmotic pressure" of seawater, which is about 30 atmospheres), the migration stops. At this pressure, the rate of migration of water molecules from seawater into fresh water is equal to the rate of migration in the opposite direction. If we increase the pressure even further, water begins to migrate in the reverse direction – from seawater into the freshwater.

A simple, short (1-1/2 minute) video illustrating the basic concept of reverse osmosis can be viewed at: https://www. youtube.com/watch?v=4RDA_B_dRQ0

Reverse Osmosis

Applied pressure

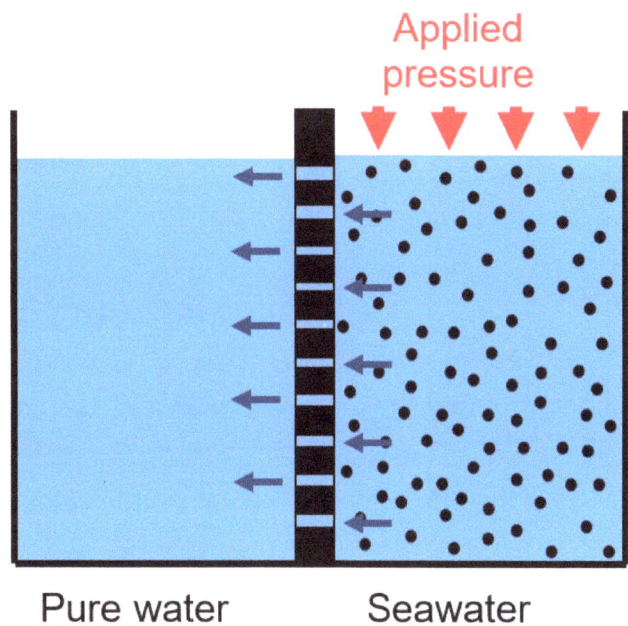

Pure water   Seawater

You might think that application of high pressure to the seawater "pushes" water molecules through the membrane pores at a faster rate, but this is not really what happens. Any water molecule in the seawater that strikes a pore will go through, whether pressure is applied or not. But, application of high pressure to the seawater reduces the rate at which water molecules migrate from the fresh water into the seawater. To pass through the membrane, water molecules must have enough energy to push against the applied pressure P. Each mole of water molecules has volume V_m, and the energy required to push each mole of water molecules against the applied pressure is PV_m. Molecules in the fresh water face an energy barrier PV_m to pass through the membrane.

However, a fraction of the water molecules are moving with enough energy to sail "uphill" against this energy barrier (just as some air molecules have enough energy to rise high up in the atmosphere). Of those water molecules in the fresh water that strike a membrane pore, the fraction with enough energy to pass through the membrane is:

$$\text{Fraction of molecules with enough energy} = e^{-P\,V_m/RT}$$

On the seawater side of the membrane, water molecules comprise the fraction X_{WATER} of the particles that wander into pores of the membrane. Of all the particles in seawater that strike membrane pores, the fraction X_{WATER} are water molecules that can pass through the membrane.

Once the applied pressure exceeds the osmotic pressure Po, then the number of water molecules passing from the seawater into freshwater begins to exceed the number of water molecules passing in the opposite direction. The condition at which this begins to happen is:

Eq (3) $$e^{-P_o\,V_m/RT} = X_{WATER}$$

When the applied pressure exceeds the osmotic pressure **Po**, then there is a net migration of water molecules out of the seawater, through the membrane, and into the fresh water.

Just as in Mechanical Vapour Recompression, there is a minimum pressure difference – and a minimum amount of work - required to transfer each mole of water from salt solution into fresh water. Each mole of water, with volume V_m, must be pushed against the osmotic pressure P_o, requiring work $P_o V_m$

Re-arranging Equation (3) gives the osmotic pressure of seawater and the work required to separate each mole of fresh water from seawater:

$$\text{Osmotic pressure } P_o = -\frac{RT}{V_m} \ln X_{WATER}$$

$$\text{Work required, } P_o V_m = -R\,T \ln X_{WATER}$$

And using the same mathematical simplification as before:

Eq (4) $$\text{Work required per mole water} = R\,T\,X_{SALT}$$

The amazing thing is that *the minimum work required to produce each litre of fresh water by reverse osmosis (given in Equation 4) is <u>exactly the same</u> as the minimum work required to produce each litre of water by mechanical vapour recompression (as given in Equation 2)!*

The minimum energy required to extract a litre of fresh water from seawater (about 3,000 Joules) doesn't arise from a limitation of the technology, but is a fundamental consequence of the 2nd Law of Thermodynamics. A solution of salt mixed with water has more molecular disorder (higher "entropy") than its separate components (water and salt). Mixing of salt with water occurs spontaneously, and is accompanied by an increase in entropy. To reverse the process, and separate water from the salt, energy must be expended. The amount of energy required (**R T X_{SALT}** per mole of water) depends upon the concentration of salt solution.

This is the same principle which determines the minimum energy required for refrigeration. A bottle of cold beer sitting on a warm kitchen table also has high molecular disorder (high "entropy"). The cold beer bottle will spontaneously come to the same temperature as its surroundings, but to reverse the process (and remove heat from the beer bottle) energy must be expended. The amount of energy required to remove each unit of heat depends upon the temperature difference and the absolute temperature.

To desalinate seawater using the minimum possible energy would require "perfect technology" which has no losses, no friction, and optimal operating conditions.

In the real world, energy requirements for desalination will always be greater than the idealised minimum.

- For one thing, as we extract water from seawater solution, the remaining salt solution becomes increasingly concentrated (the mole fraction of salt X_{SALT} increases), so more pressure must be applied [note 6]. We can minimise this effect by circulating large volumes of seawater through our reverse osmosis unit, and extracting only a small fraction of the water content, but this requires additional pumping energy. We can only hope to achieve the best trade-off between these two losses. Usually, about half the water content of the seawater is extracted.

- Secondly, if we apply only the osmotic pressure – or just slightly above - to the seawater solution, the production of fresh water will be incredibly slow. To produce fresh water at a practical rate, we need to apply pressures that are significantly above the osmotic pressure of seawater.

With current technology, the most efficient reverse osmosis units require about 10,000 Joules to produce each litre of fresh water, about three times as much mechanical pumping energy as the theoretical minimum energy requirement. The energy requirement for desalination could be provided by renewable energy (solar or wind), fossil fuels or even by human muscle power.

Small, hand-powered reverse osmosis units are produced as emergency equipment for boat owners. One such unit is reportedly able to produce 4-5 litres of potable water per hour. Pumping by hand for long periods would be quite tedious, and would not produce enough water for bathing or washing clothes, but it could keep alive a few sailors adrift in a lifeboat.

The first vegetable farm in which water is provided by solar-powered desalination was established in a desert environment near Port Augusta, South Australia. The farm is expected to produce 15,000 tonnes of tomatoes per year.[Note 7]

Using current reverse osmosis desalination technology, electrical energy generated by one kilowatt of solar panels could produce about 600 kilolitres of fresh water per year. About 2-3 kilowatts of solar panels would be needed to produce one tonne of (dried) rice each year, and 20 kilowatts of solar panels would be required for each hectare of rice cultivated.

Usually, the energy source for desalination plants is electricity produced by burning fossil fuels (coal, natural gas, diesel fuel, etc). Since generating electricity from fuel is only about 35% efficient, about 30,000 Joules of fuel energy is consumed per litre of water produced. More than a kilogram of coal would be consumed at a power station for each thousand litres of fresh water produced by desalination. If the desalinated seawater were used to irrigate rice crops, at least one-and-a-half kilograms of coal would be consumed for each kilogram of rice produced.

A 6-1/2 minute video describing reverse osmosis can be viewed at:
https://www.youtube.com/watch?v=mZ7bgkFgqJQ

A four-minute video giving a good description of the technology can be viewed at:
www.youtube.com/watch?v=aVdWqbpbv_Y

To extract freshwater from seawater, we must expend energy. But the process is reversible. It is also possible to utilise the "entropy of mixing" of freshwater and seawater to produces an *energy output*. An "osmotic engine" could produce energy by mixing freshwater and seawater. Researchers have proposed that such an osmotic power plant be constructed at a river mouth, where fresh water flows out and mixes with the ocean. The fresh water and seawater streams would flow on opposite sides of a semi-permeable membrane, generating the osmotic pressure within the seawater stream. Pressurised seawater could then drive a hydraulic motor, producing work to drive a generator and produce electricity.

Operating a reverse osmosis unit in reverse, as an "osmotic engine", could produce a maximum of 3,000 Joules for each mole of fresh water that is mixed with seawater. This is exactly the same as the minimum work required to separate one mole of fresh water from seawater.

Reverse osmosis has been around for more than 50 years, and mechanical vapour recompression (MVR) has been around for even longer. You might think that, in future, a new type of desalination technology will be developed that will be more efficient than reverse osmosis or MVR could possibly be. This super-duper desalination technology, you might argue, would require less energy to produce fresh water than the amount given by Equation 4. But you would be wrong! We can conduct another thought experiment to show that no future desalination technology could ever be more efficient that an ideally-efficient reverse osmosis unit.

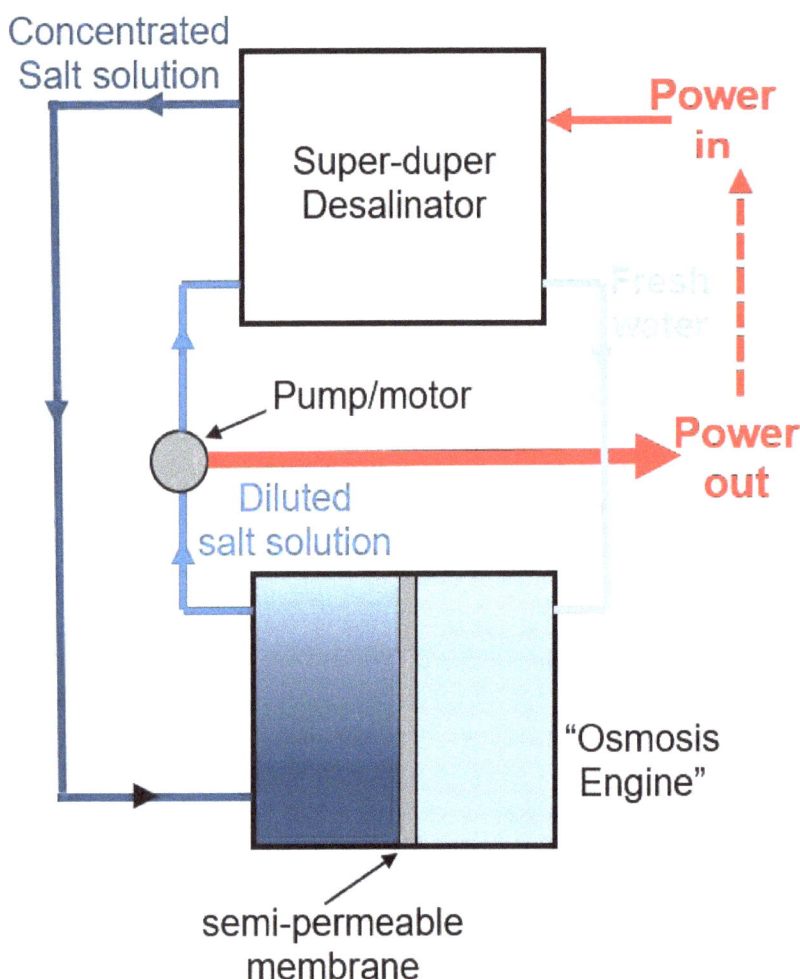

We could operate our best possible reverse osmosis system as an "osmotic engine" producing the maximum possible work output ($RT \ln X_{SALT}$ for each litre of water that is mixed with seawater). The osmotic engine mixes fresh water with a saltwater solution, producing a diluted seawater stream. We return the diluted seawater stream to our super-duper desalination technology, which extracts the fresh water that has been mixed, and regenerates saltwater at the original concentration. In this way, the fresh water and saltwater solutions are continuously cycled around and around, being mixed and unmixed. We use the work output of the "osmotic engine" to provide the energy needed to operate the super-duper desalinator.

If the super-duper desalination unit requires less energy than our "perfect" reverse osmosis unit operating in reverse, then each time fresh water and salt solution are cycled through the system, the work produced by the "osmotic engine" output would be more than the work needed to operate the super-duper desalinator. The system would produce a net work output, with no energy input. This would violate the Second Law of Thermodynamics, which we could be reasonably confident would never happen. This means that *no desalination technology could ever be developed that requires less energy than the best possible reverse osmosis or mechanical vapour recompression unit. No desalination technology will require less energy than the amount given by Equation 4*.

Reverse osmosis and mechanical vapour recompression are amazing technologies that have been greatly improved by technological advances over the last few decades, and will probably be further improved in the years ahead. But it is not a "magic bullet" or a "Get out of jail, free" card that will avoid impending challenges of supplying enormous amounts of water to feed a growing world population. Desalination of water is financially costly and energy-intensive. Even with further technological improvement, desalination of seawater will always have a significant energy cost.

References and notes

1. US Geological Survey, Ground water depletion across the nation, Nov 2003, https://pubs. usgs.gov/fs/fs-103-03/JBartolinoFS(2.13.04).pdf

2. Yoshihide Wada et al, A worldwide view of groundwater depletion, December 2010

    https://www.researchgate.net/publication/48326130_A_worldwide_view_of_groundwater_depletion

3. Kay Daigle, Death in the water, Scientific American, January 2016, p. 38-47

4. The table of irrigated crops is based on data published in the Australian Newspaper (9 May 2018, Inquirer section, p. 25).

5. If you picture the Earth as a sphere with a radius **R** of 6,380 kilometres, you can see that the Earth has a cross-section area of πR^2 intercepting radiation from the sun. The intensity of sunlight reaching the Earth's surface on a clear day happens to be 1,000 watts for each square metre of area perpendicular to the sun's rays. So, the total solar radiation striking the Earth's surface is (1,000 watts/m^2) [(π (6.38 X 10^6 metres)^2] = 128 X 10^{15} watts.

    About 40% of incident solar radiation is reflected into space, and the remaining 60% (76 X 10^{15} watts) is absorbed and converted into heat. This amount of heat can evaporate 34,000 million litres of water each second. After being evaporated, water vapour condenses when humid air is cooled at night or rises to higher elevation. Condensation may occur hundreds or thousands of kilometres away from where it evaporated, and days later. Over a long period of time, and across the Earth's surface, water vapour is condensing into rainfall at the same rate as it is evaporating.

    If we spread the total amount of water that can evaporate (and then condense as rainfall) across the entire surface of the Earth, we calculate an annual average global rainfall of 0.98 metres per year. This is about 15% more than the reported global average rainfall of 0.86 metres/year, indicating that Earth is a very effective solar-powered distillation machine.

6. As we extract water, the salt concentration of seawater solution increases until reaches about ten times its initial concentration. At this point, the salt solution becomes saturated and solid salt crystals begin to crystallise from the solution.

7. "First farm to grow veg in a desert using only sun and seawater". New Scientist, 6 October 2016. https://www.newscientist.com/article/2108296-first-farm-to-grow-veg-in-a-desert-using-only-sun-and-seawater/

11. Extracting water from the air

Any nation or region that has access to the ocean could potentially obtain its fresh water needs by desalinating seawater (if the community is prepared to pay the financial and energy cost). But not all regions have access to the oceans. Some areas are too far from the coast for fresh water to be transported by pipeline or other means. However, everywhere on the Earth's surface provides access to the atmosphere, which contains water vapour.

Depending upon the local climate and conditions, water vapour can comprise as much as 5% of the molecules in the air. This occurs in hot humid conditions (generally encountered in steamy jungles or saunas) when the relative humidity approaches 100%. In this case, the gas pressure arising from water vapour (the "partial pressure" of water vapour) approaches the equilibrium vapour pressure at that temperature.

Remember that the amount of water vapour in saturated air depends dramatically on the temperature, roughly doubling for each 10°C rise in temperature. The relative humidity **H** tells us the amount of water vapour contained in the air, *relative to the amount that the air could hold if it was saturated*. Here is a graph showing the percentage of water vapour in saturated air, and in air at 50% relative humidity (that is, containing 50% as much water as saturated air).

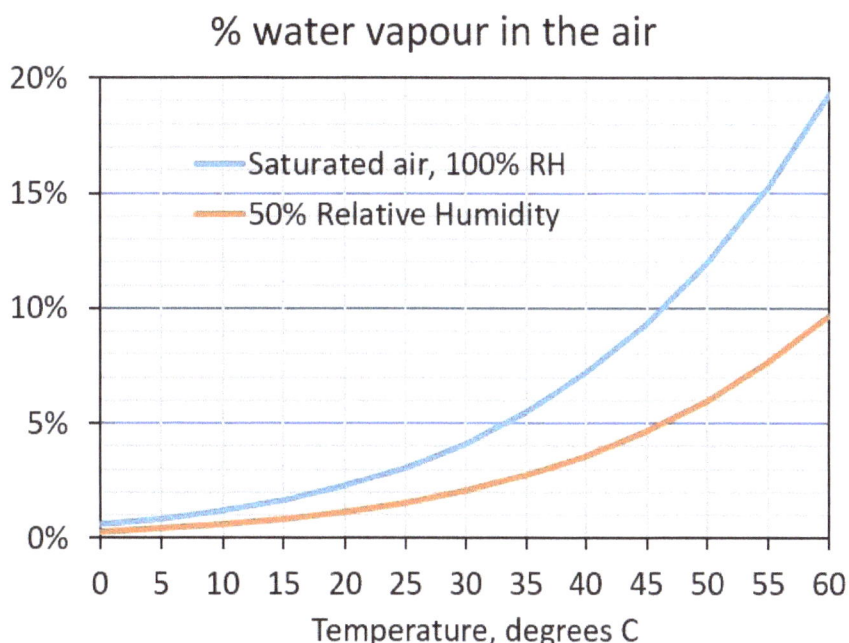

% water vapour in the air

In very few places does the relative humidity often approach 100%. In most coastal cities, like Brisbane, the relative humidity throughout the year is typically about 50%. On summer days, we would generally regard conditions as uncomfortable and oppressive when the humidity reaches 70%. If you did any physical exertion, your sweat would be slow to evaporate, and your shirt would probably become soaked with sweat. On the other hand, in arid inland Australia, the relative humidity might only be 20%.

Since the air always contains water vapour, it is possible to extract liquid water from the air, anywhere on Earth. In fact, this happens all the time. You have probably noticed droplets of water forming on a cold bottle of beer or soft drink after it is removed from the refrigerator on a warm, humid day. You have probably also seen water dribbling from air conditioning systems, especially on humid days.

Consider a typical summer day in Brisbane, when the air temperature is about 30°C and the relative humidity is 50%. In this case, as can be seen on our graph, the air contains about 2% water vapour. When we cool the air, its relative humidity increases, and the air becomes fully saturated (100% relative humidity) at about 18°C. If the air is cooled any further, liquid water begins to condense.

% water vapour in the air

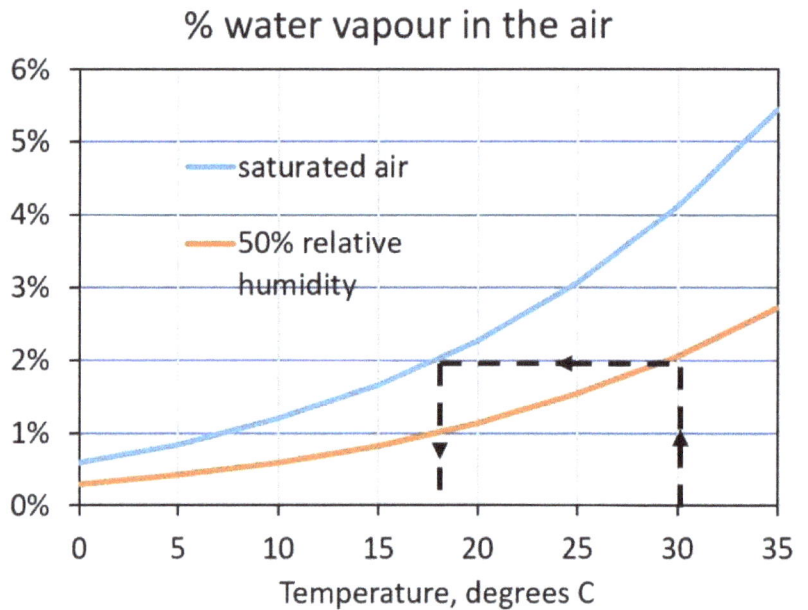

Temperature, degrees C

The concept of extracting water by cooling the air is applied in some commercial systems. These incorporate a refrigeration system to cool air from the outside air temperature T_{AIR} to below its "dew point" temperature T_{COLD}. Droplets of liquid water condense and are collected. The remaining cold air may then be passed through a heat exchanger or regenerator to pre-cool incoming air. In principle, by using such a heat exchanger, we can recover and recycle all of the refrigeration energy that was used to cool the air to its dew point. *But, to condense the cool saturated vapour to liquid, we still need to remove its heat of vaporisation at temperature T_{COLD}.*

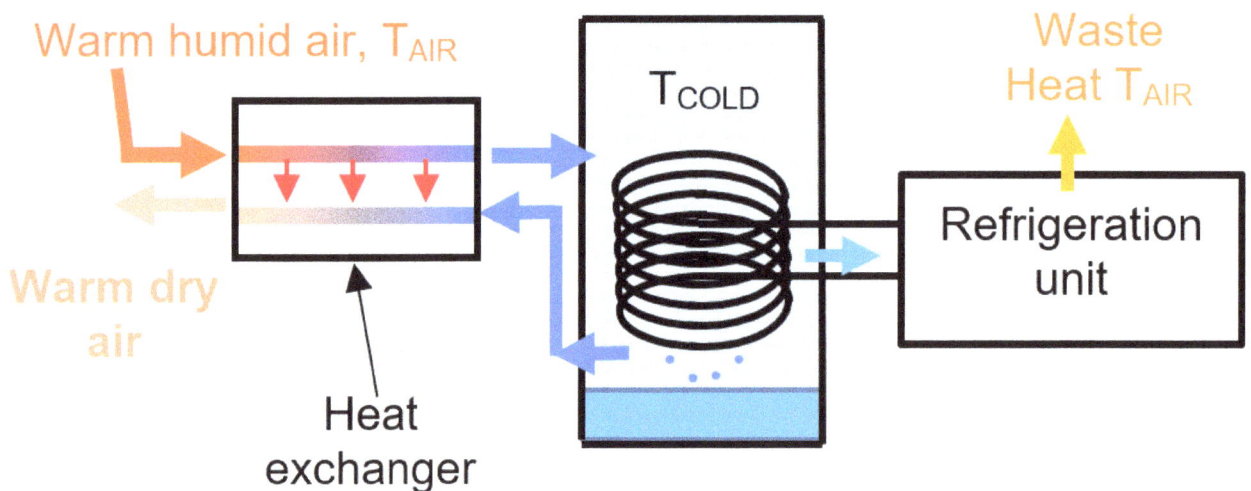

Since the vapour pressure of water varies exponentially with temperature, the reduction in temperature $T_{AIR} - T_{COLD}$ needed to bring the air to its dew point can readily be calculated. As we might expect, the temperature reduction needed for the air to become saturated with water vapour depends directly on the humidity of the air. It turns out [see Note 1], that the reduction in temperature is:

$$\text{Temperature reduction } T_{AIR} - T_{COLD} = \frac{R \, T_{AIR} \, T_{Cold} \ln (H)}{E_{vap}}$$

This temperature reduction determines the maximum possible coefficient-of-performance of the refrigeration unit, and therefore, the minimum work input required to remove the heat of vaporisation at T_{COLD}.

Recall that the maximum Coefficient-of-performance of a refrigerator is:

$$\text{Maximum Coefficient of Performance} = \frac{T_{COLD}}{T_{AIR} - T_{COLD}}$$

And, therefore, the minimum work required to condense one mole of water by removing its heat of vaporisation E_{vap} is:

$$\text{Minimum work to condense one mole of water} = \text{Evap} \left[\frac{T_{AIR} - T_{COLD}}{T_{COLD}} \right]$$

Finally, substituting for the value of $T_{AIR} - T_{COLD}$ gives simply:

Equation (1) $\text{Minimum work} = R\ T_{AIR}\ \ln\left[\frac{1}{H}\right]$ per mole of water extracted

At an air temperature **T** of 27°C (300° absolute) and relative humidity of 50% (H = 0.5), the minimum work required is 1,700 Joules per mole, or about 95,000 Joules per litre. ***This is the absolute minimum energy that would be required with perfect technology*** (no friction or losses, perfect operating conditions). This is ***thirty times more*** than the minimum energy required to extract fresh water from seawater by desalination.

In some situations, where water is essential for survival, the high energy consumption could be an acceptable price to pay. Using perfect technology (using the minimum energy consumption given by Equation (1), one kilowatt of solar panels could produce enough energy to extract 180 litres of water per day from the air at 50% humidity. However, existing water-from-air technology requires far more than the minimum energy requirement, and it is hard to imagine that this technology could ever be a viable means of producing water for agriculture.

Why does extracting water from the air require so much more energy than to extract water from seawater? In air at 50% relative humidity, water molecules have only half the concentration that would be present at equilibrium. In seawater, the water molecules that we want to extract comprise 98% of the molecules/ions that are present. To separate water molecules from air requires that we concentrate them to a much greater extent – that is, that we greatly reduce the "molecular disorder" (or "entropy") of the water molecules. To do this, energy must be expended.

There is another method of extracting water from the air, of which I first became aware some 40 years ago. At that time, I was learning to drive large army trucks for the US Army reserve. These trucks had power brakes based on compressed air. An engine-driven pump compressed air to several atmospheres pressure, and the compressed air was stored in a tank mounted on the truck chassis.

When air is compressed, each of the gas components in the air (including water vapour) is compressed to higher pressure. If the "partial pressure of water vapour" (the pressure of water vapour molecules in the gas) exceeds the equilibrium vapour pressure of water at that temperature, liquid water condenses. In the case of the compressed air tank of a truck, condensation of liquid water can be problem if the truck is parked in frigid conditions. Condensed water freezes into ice that can block the air lines (leaving the truck with no brakes). Particularly in cold winter climates, it is important that collected water is drained, and many trucks now have automatic drain valves.

The same principle can be used to extract liquid water from the air. We start by drawing in air at atmospheric pressure, whose partial pressure of water vapour is equal to the relative humidity **H** multiplied by the equilibrium vapour pressure at ambient temperature P_{AIR}. Then, we compress the air, while keeping the gas at constant temperature (as this minimises the work required to compress the gas). Once the partial pressure of water vapour reaches P_{AIR}, the gas is saturated with water vapour. As the volume of the air is further reduced, the partial pressure of water vapour remains constant at the equilibrium vapour pressure P_{AIR}, and droplets of liquid water condense. The droplets of liquid water can be collected and separated from the gas stream.

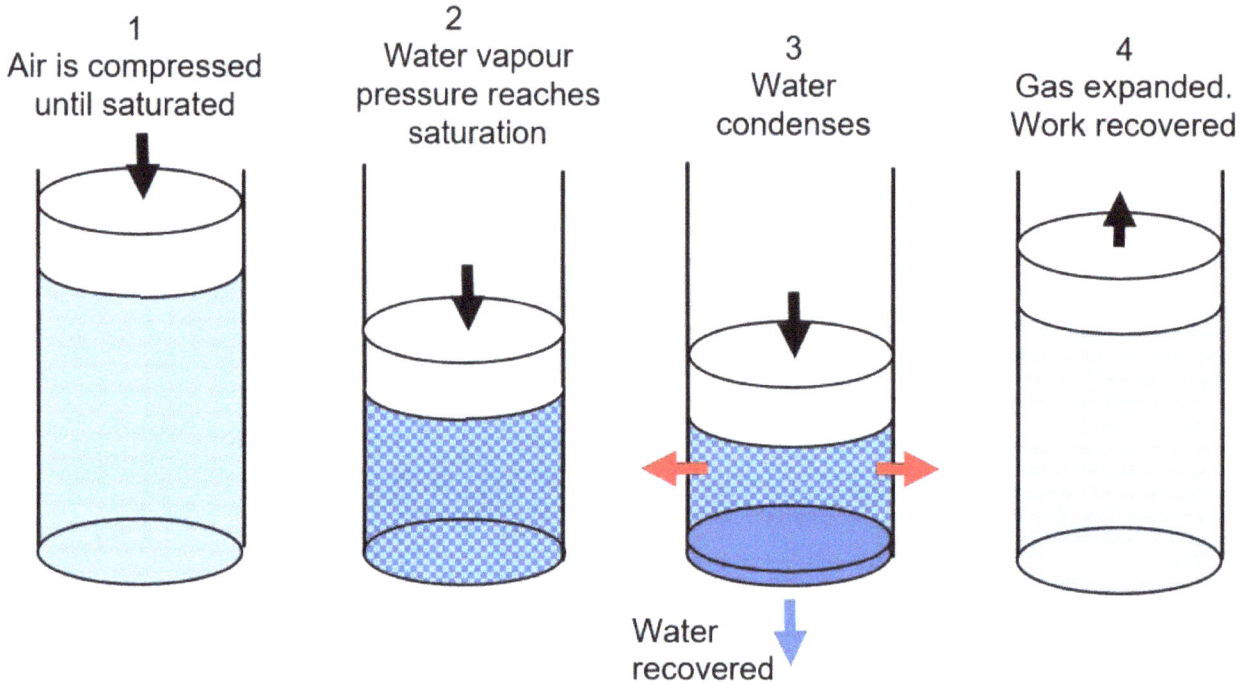

1	2	3	4
Air is compressed until saturated	Water vapour pressure reaches saturation	Water condenses	Gas expanded. Work recovered

Water recovered

Once some water vapour condenses as liquid and is separated, the remaining air can be expanded back to its original atmospheric pressure. If we have an ideal compressor/expander, the work used to compress the air can be recovered by expanding the remaining air back atmospheric pressure. But even with an ideal, frictionless compressor, the air will not return to its original volume - because water vapour that was originally present has been condensed and removed from the gas. All the work required to compress the gas can be recovered – except for the work required to compress the water vapour that was removed. For this water vapour to reach saturation, its pressure had to be increased from $P_{AIR}H$ to P_{AIR}, or by a factor of **1/H**.

In other works, the pressure of the water vapour must be increased by a factor of **1/H**. So, for air at 50% humidity, water droplets begin to condense when the volume of the air is reduced to half.

We have previously seen that the work required to compress air, or any gas, is related to its initial and final pressure as follows [Note 2]:

$$\text{Work to compress a gas} = n\,R\,T_{AIR}\,\ln\left[\frac{P_{FINAL}}{P_{INITIAL}}\right]$$

Where **n** is the number of moles of gas
R is the Universal Gas Constant
T_{AIR} is the absolute temperature of the gas
P_{FINAL} is the final pressure of the gas (after compression)
$P_{INITIAL}$ is the initial pressure of the gas (before compression)

Since the ratio of the final and initial pressures is **1/H**, the minimum work required for each mole of water extracted is:

$$\text{Minimum work} = R\ T_{AIR}\ \ln\left[\frac{1}{H}\right] \quad \text{per mole of water extracted}$$

This is exactly the same result derived for Equation (1)!

This leads to the conclusion that extracting water from the air is inherently far more energy-intensive than desalination of seawater. Extracting water from air would be more most desirable and valuable in inland locations, where transport of water from the coast (generally through pipelines) would likely be prohibitively expensive and energy-intensive. However, at least in Australia, inland locations tend to have lower relative humidity, with even higher energy requirements to extract water from the drier air.

The energy requirement for extracting water from the air increases dramatically in drier climates. In an arid climate, with relative humidity of 25%, the minimum energy requirement is twice as great as in typical coastal environments with humidity of about 50%.

In central Australia, daytime humidity may be as low as 10% in summer. At relative humidity of 10%, the minimum energy requirement is more than three times as much as in typical coastal environments with humidity of about 50%.

Minimum energy to extract water from air at 30°C

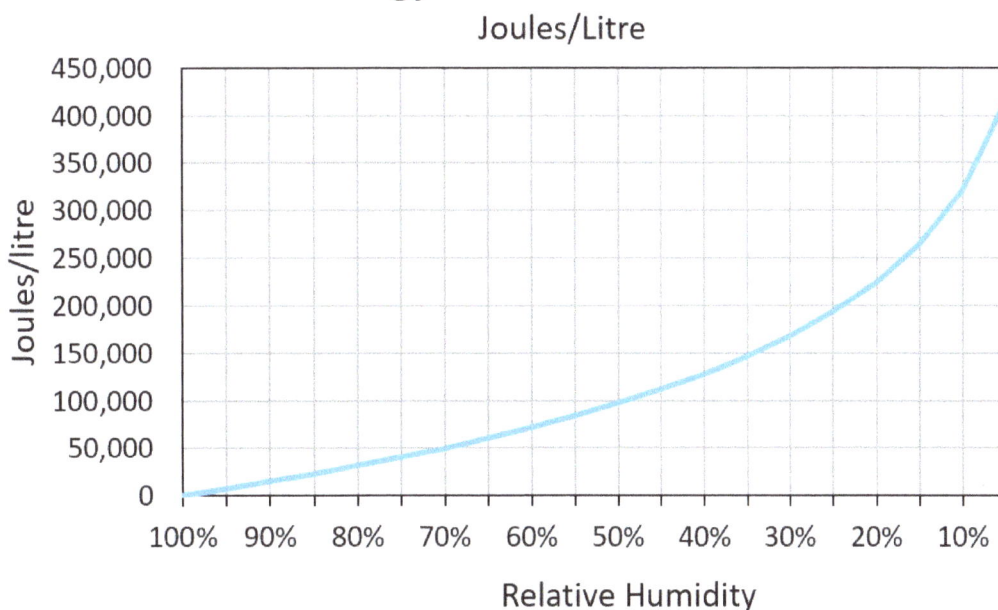

The actual energy consumed in extracting water from air will always be greater than the theoretical minimum given by Equation (1), partly due to the same reasons that the energy required for reverse osmosis will always be greater than the theoretical minimum. In fact, actual energy consumption of water-from-air systems that are currently available seem to be much higher than the theoretical minimum. The energy-efficiency of these systems is hard to judge, as the promotional material on their websites rarely specifies the humidity at which the stated water production and energy consumption is measured.

One commercial water-from-air system is stated to produce 378 litres/day of water with power consumption of 4.2 kilowatts at 75% humidity. This corresponds to an energy consumption of 960,000 Joules for each litre of water produced (25 times the theoretical energy consumption). This is consistent with the statement on the company's website that their diesel-powered system can produce 10-14 litres of water for each litre of diesel fuel. This might be acceptable for an emergency water supply, but the cost of diesel fuel alone (12 cents for each litre of water) would be thirty times more than Brisbane residents are currently charged for their water consumption.

Another company has undertaken conceptual design of a utility-scale water extraction plant which, if constructed, would be the largest water-from-air plant ever built. The plant is intended to produce 400,000 litres of water per day (at unspecified humidity), requiring a continuous 6 megawatts of power, produced by solar panels and wind turbines. This is an energy input of 1.3 megajoules for each litre of water produced [Note 3]. Solar panels that would provide the energy for such a plant might cost perhaps $25 million (at $1/watt), and finance charges for the solar panels and battery storage alone would correspond to about 1.5 cents for each litre of water produced. It is conceivable that such a plant could provide a water supply for a remote community with 1,000 residences, although construction and operational costs of the plant would be prohibitive unless the community had access to very large financial resources or subsidies.

In theoretical and practical terms, extracting water from the air is far more energy-intensive than desalinating seawater. In coastal areas, where the relative humidity would typically be about 50%, the minimum energy required to extract water from the air is 30 times more than to desalinate seawater. One would expect that water extracted from air would be particularly valuable in inland locations, where transport of water from the coast (generally through pipelines) would be expensive and energy-intensive. However, inland locations tend to have lower relative humidity, with even higher energy requirements to extract water from the drier air.

However, with a supply of sufficient energy, this technology would make it possible to grow food crops in arid inland regions that are entirely unsuited for conventional agriculture. This could even be possible in desert regions, where the humidity might typically be 20%. Rather than recover water directly from the air (at very low humidity), a more efficient option would be to recover water from air after it passes through an enclosed greenhouse, gaining moisture from crop plants as they transpire. Carbon dioxide needed by the plants would be provided by drawing low-humidity outside air into the greenhouse. As this air passes over the plants, it would gain moisture by transpiration, perhaps reaching 60% humidity. Before being released back to the atmosphere, this humidified air would be passed through a water recovery unit, which would extract most of the water vapour, and reduce the humidity back to its initial level (of, say, 20%). Liquid water that is recovered would be recycled back to the roots of the plants. In this way, all the water that the plants lose by transpiration is recycled back to the plants.

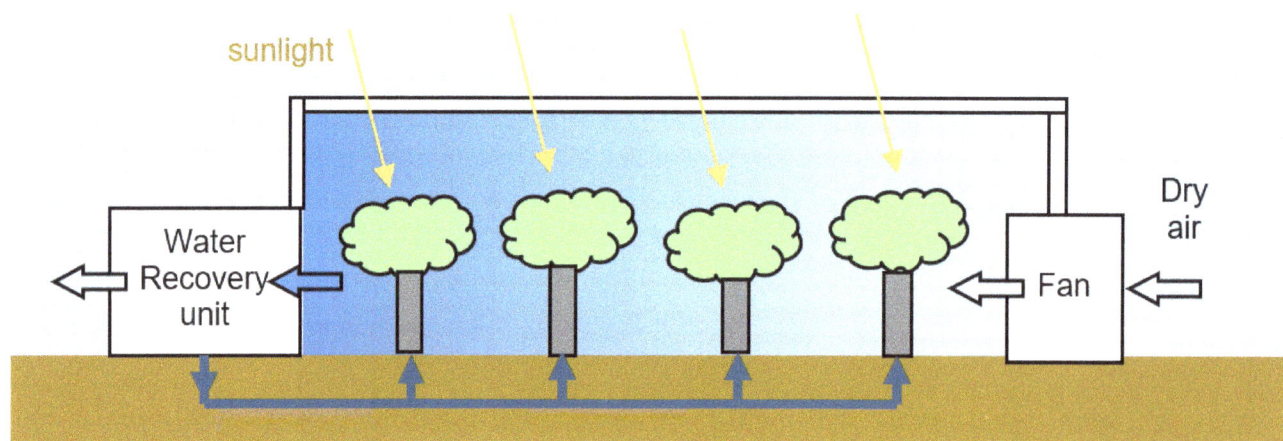

The minimum amount of energy required to extract and recover the water would depend upon the final humidity of the air after passing over the plants and the initial humidity of the outside air. I have calculated the theoretical minimum amount of energy required for each litre of water recovered [See note 4]. I plotted the energy requirements versus the outside air humidity for three scenarios (in which the final humidity, after passing over the plants, is 50%, 60% and 70%).

Here are graphs of my results:

Energy to extract water from air, Joules/litre at 25°C

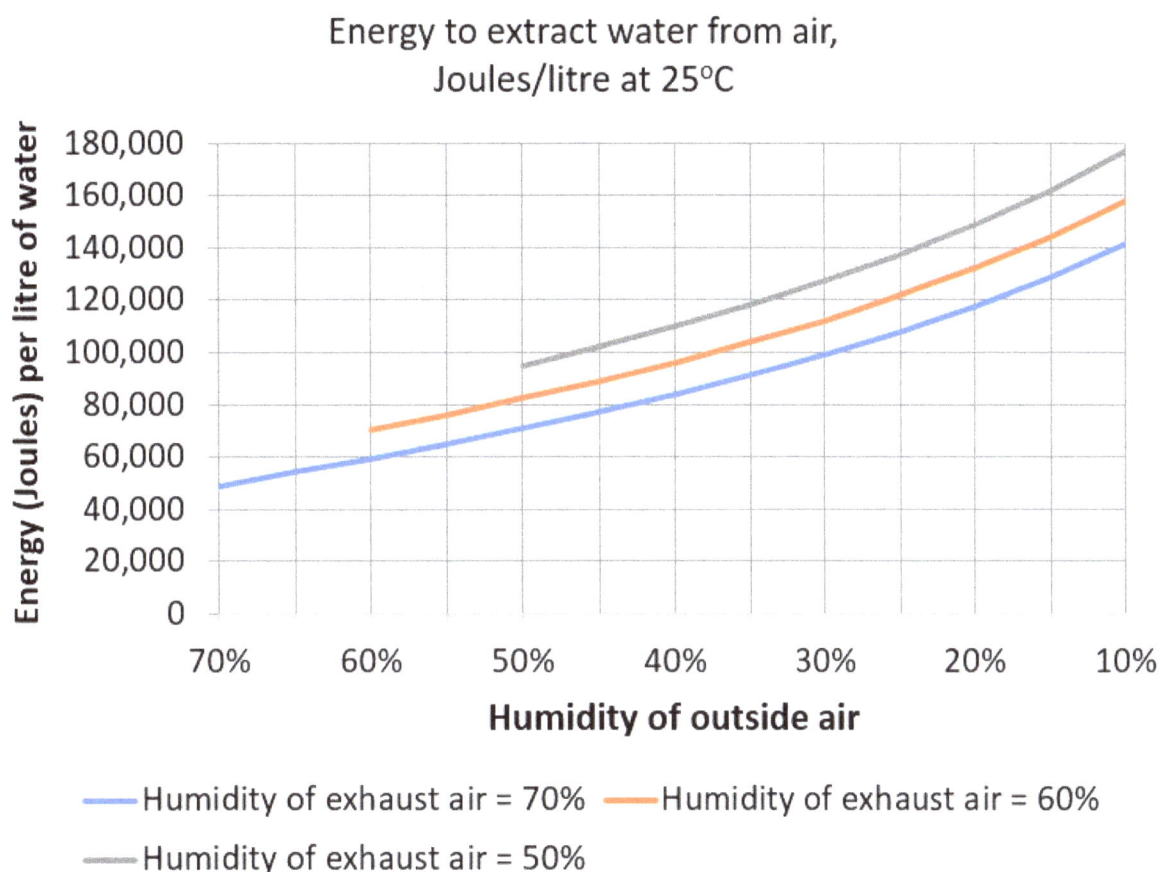

Legend:
— Humidity of exhaust air = 70% — Humidity of exhaust air = 60%
— Humidity of exhaust air = 50%

In the example given earlier, with an outside air humidity of 20% and a final humidity of 60% (orange graph), the minimum energy required to recover each litre of water is 132,000 Joules. Let's say that we choose to grow almonds, using existing agricultural practices, which require about 14 million litres of water per hectare. Then, the minimum energy requirement to operate the water recovery unit would be 1,850 gigajoules for each hectare of the almond crop. This much energy could be produced by 12,000 square metres, or 1.2 hectares, of solar panels. To provide sufficient power to operate the water recovery unit would require at least 2.5 megawatts of solar panels, covering an area larger than the crop. The cost of the solar power plant alone would likely be at least $2 million for each hectare of almond crop, yielding 2 tonnes of almonds per year. Simply to pay back the capital cost of the solar power plant over 20 years would add $50 per kilogram to the cost of the almonds.

It might be possible to do better than this. If crops are being grown in such a very expensive climate-controlled greenhouse, it would make economic sense to use the most water-efficient agricultural practices that are available. Such practices might significantly reduce water loss by transpiration.

While water extraction from air will likely be prohibitively expensive and impractical for widespread application on Earth, it may be necessary for proposed human habitation on Mars or the moon (or on spacecraft travelling for long periods). Such habitation will require a fully-enclosed environment with its own atmosphere. If a human settlement on another planet is to be sustainable and permanent, it must recycle water, oxygen, carbon dioxide and nutrients, and grow its own food supply. Control of water and humidity would be critical. Water lost through transpiration by food crops (or exhaled by the astronauts) would need to be extracted from the air to be re-used, and to maintain humidity within a range that is comfortable and healthy for human inhabitants. Water vapour could be condensed through either a refrigeration of gas compression/expansion system, or by utilising the day-night temperature fluctuations on the planet/moon. Electrical power required to operate the system could be provided by solar panels or (more likely I suspect) by a small nuclear power reactor.

Consider for example, an enclosed human settlement on Mars. To sustain itself, the settlement would need to be essentially self-contained and isolated from the surrounding Martian atmosphere. The settlement would need to grow enough food crops to feed its population. Growing food plants would also enable the atmosphere inside the settlement to be a "closed loop", with the plants absorbing carbon dioxide exhaled by the human occupants and producing oxygen. However, to maintain a "closed loop" water cycle (and maintain a humidity that is healthy for plants and humans), it would be necessary to recover water vapour transpired by plants, and recycle fresh liquid water to the plants.

It may be possible to condense water vapour by passing air within the enclosed settlement through a heat exchanger, transferring heat to the thin, cold atmosphere of Mars. Alternatively, water vapour transpired by plants could be recovered in a water recovery unit powered by a small ten megawatt nuclear reactor (about the size of the reactors on nuclear submarines). To grow sufficient crops to feed the settlement, about 1,000 litres of fresh water will be needed each day per person. If the water vapour recovery system is as efficient as it could possibly be (as given by Equation 1), the power output of a 10 MW reactor could recover and recycle enough water to feed as many as 10,000 people. Allowing for inevitable losses and inefficiencies, it might be feasible to support a population of 1,000 people.

With its equipment and key construction materials supplied from Earth, the cost of building such a human settlement would be enormous. Advocates of human settlement on other planets argue that, once established, such a settlement could operate without regular re-supply from Earth. However, it is questionable whether this could be achieved. The sophisticated equipment on which the settlement will depend will eventually wear out, and the reactor would eventually need to be refuelled. It remains to be seen whether a community of, say, 1,000 men, women and children could have the capability to manufacture replacement parts and materials needed to sustain such a complex self-contained life support system.

Notes

1. Normally, the air is not saturated with water vapour. At air temperature T_{AIR}, the pressure of water vapour is a fraction of the equilibrium vapour pressure P_{AIR}. This fraction is the humidity of the air, H. If we take air at normal air temperature T_{AIR}, the pressure of water vapour is $P_{AIR}H$. As we cool the air, the vapour pressure of water remains the same, but the equilibrium vapour pressure reduces, and the humidity increases. At the "dew point" temperature T_{COLD}, the air is fully saturated with water vapour and the humidity is 100%.

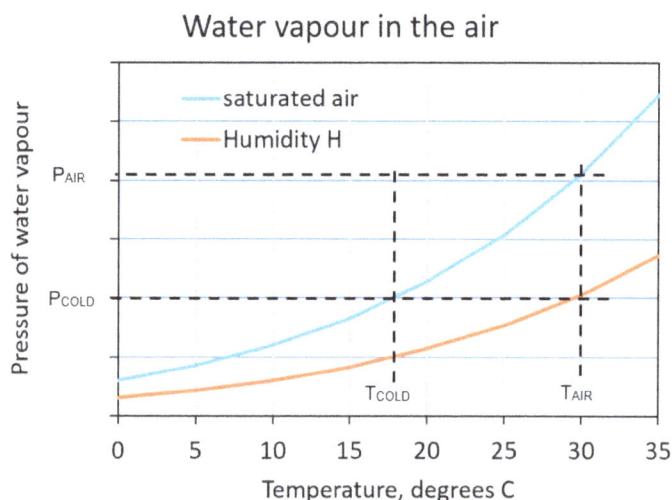

Water vapour in the air

The equilibrium vapour pressure of water at the dew point P_{COLD} is equal to the equilibrium vapour pressure P_{AIR} at air temperature multiplied by the humidity H. In other words:

$$P_{COLD} = P_{AIR} H \qquad \text{So,} \quad \frac{P_{AIR}}{P_{COLD}} = \frac{1}{H}$$

The equilibrium vapour pressure of water, or any volatile liquid, varies exponentially with temperature. The vapour pressures P_{AIR} and P_{COLD} at temperatures T_{AIR} and T_{COLD} are related by:

$$Ln\left[\frac{P_{AIR}}{P_{COLD}}\right] = \frac{E_{vap}}{R}\left[\frac{1}{T_{COLD}} - \frac{1}{T_{AIR}}\right]$$

Combining these two equations and re-arranging gives the temperature difference $T_{AIR} - T_{COLD}$ that the air must be cooled to reach the dew point:

$$T_{AIR} - T_{COLD} = \frac{R\, T_{AIR}\, T_{COLD}\, ln(1/H)}{E_{vap}}$$

2. To compress a gas using the **minimum** amount of work, the gas should be kept at constant temperature T. This called an "isothermal compression", and the work required to compress n moles of gas from its initial pressure $P_{Initial}$ to its final pressure P_{Final} is given by:

$$\text{Work to compress a gas at constant temperature} = n\,R\,T\,Ln\left[\frac{P_{Final}}{P_{Initial}}\right]$$

3. The refrigeration system for the proposed plant requires a continuous power input of 6 megawatts, or 6,000 kilowatts. The energy requirement over a 24-hour period is:
   (6,000 kilowatts)(24 hours) = 144,000 kilowatt-hours

   Converting this to units of megajoules:
   (144,000 kilowatt-hours)(3.6 MJ/kWh) – 518,00 megajoules

   The system is intended to produce 400,000 litres of fresh water per day, so the energy requirement per litre of water is:
   $$\frac{518,000 \text{ MJ}}{400,000 \text{ litres}} = 1.3 \text{ MJ/litre}$$

4. Consider the extraction and recovery of water from air at temperature T and initial humidity H_1, with the final humidity of the air reduced to H_2. I have calculated that absolute minimum energy (Joules) required for each mole of water recovered is:

$$\text{Minimum energy per mole of water recovered} = RT\left[1 - \frac{H_1 \ln(H_1) - H_2 \ln(H_2)}{H_1 - H_2}\right]$$

To get the minimum energy per litre of water, multiply the energy/mole by 55.5.

12. Melting of metals

For past millennia, mankind's technological progress has been based upon the ability to produce and shape metal tools, weapons and implements. About seven thousand years ago, people began to produce bronze. Later, this was followed by iron. This started the "age of metals", which has not ended. The modern world would be inconceivable without the use of metal alloys in high-rise buildings, bridges and other large structures; cars, trucks, ships and planes. Metal fabrication has relied on heating and melting the metal - initially using charcoal or coal, and later using natural gas or electricity to provide the energy required.

Until the industrial revolution, the technology for smelting and shaping of bronze and iron was crude, and small forges could not develop sufficiently high temperatures to melt the metal. Rather, the metal was heated near its melting point (so that it was soft, but still solid) and then hammered into shape. This technology is still used today, with steel billets being heated above 1,000°C, and then extruded and "hot-rolled" into plates and rods using high pressure rams or rollers.

Here is a 3 - minute video showing steel billets being hot-rolled:
https://www.youtube.com/watch?v=6xnKmt_gsLs

Once metals could be heated beyond their melting point, it was possible to pour the molten metal into moulds and produce much more complicated and intricate shapes.

When metal implements are broken, damaged or no longer needed, the metal can be remelted and recast into new implements, and this process can be repeated an unlimited number of times. Indeed, before the era of mass industrial production, metals were difficult and expensive to produce, and blacksmiths and metal workers made a living from refashioning metal implements. People have been recycling bronze and iron implements for thousands of years:

> " . . and they shall beat their swords into ploughshares, and their spears into pruning hooks"   The Book of Isaah, 800-600 BC

Even in our current society, with a "throw-away" mentality, most steel and aluminium is recycled into new products. Far less energy and resources are needed to heat, melt and recast existing steel or aluminium products than to produce new metal from ore.

Here are several videos showing various techniques for casting molten metal:
* Video, Lost wax metal casting, 4-1/2 minutes

  https://www.youtube.com/watch?v=33p0Nih6YkY
* Casting of cast iron cookware, mass production, 5 minutes

  https://www.youtube.com/watch?v=srlEy4z_hzY
* A video animation of aluminium die casting process, 1 minute

  https://www.youtube.com/watch?v=Pj_mjjUQad8

To cast liquid metal, we first need to heat it to its melting point. Heating can be done by burning coal or natural gas, or with electrical energy. As heat is added, the temperature of the metal rises – until it reaches the melting point. Then, as we keep adding heat, more and more of the metal melts, but its temperature remains fixed at the melting point. It may seem extraordinary that the temperature remains constant while we continue to add heat. The additional heat goes entirely into bringing about the transition from solid to liquid metal, a process that we call "melting" (or, for some reason, is technically called "fusion"). The heat required to melt one mole of a solid material is called its "molar heat of fusion", which we'll give the symbol E_m. Only after all of the metal has melted into liquid, and we continue to add heat, its temperature rises above the melting point.

Heat required for metal melting

All solids are held together by attractive forces between their constituent atoms, molecules or ions. Depending upon the particular substance, these attractive forces can be very weak, as is the case for inert gas elements (helium, neon, argon, krypton, xenon and radon). Accordingly, the melting point of helium is only one degree above absolute zero, and neon melts at 25° Kelvin. On the other hand, attractive forces between metal atoms are much stronger, as are interactions between positively and negatively-charged ions in salts. Metals melt at temperatures ranging between 245 degrees (for mercury [Note 1]) and 3,400 degrees (for tungsten) above absolute zero.

Attractive forces between atoms or molecules act over very short range, tending to pull the atoms into a highly ordered crystal structure in which each atom interacts with the maximum number of surrounding atoms. If we could see the atoms in a crystal, they would be arranged in a highly-ordered regular pattern. We can imagine a crystal structure being like a formation of well-disciplined soldiers.

In the case of spherical atoms (that is, atoms whose electron orbitals have spherical symmetry), the question arises as to how the atoms can best be stacked to achieve the closest packing,

Source: http://www.chinadaily.com.cn/photo/2013-10/17/content_17038930_5.htm

so that each atom has the maximum number of nearest neighbours. This problem confronted the Royal Navy in the 16th century, when ship's crews needed to pack the maximum number of cannonballs into the smallest storage volume in their ship's hold. The solution which gives the closest stacking of spheres is to arrange the cannon balls in layers with the spheres in a hexagonal pattern. Each layer is shifted relative to the layer below so that each atom lies above the intersection of three atoms in the underlying layer. [Note 2]

This arrangement is very energetically favourable. Each atom is directly in contact, and attracted to, six atoms in its own layer, three atoms in the layer below, and three atoms in the layer above. In total, each atom is attracted to 12 nearest neighbours (or, as crystallographers would say, has a "coordination number" of 12).

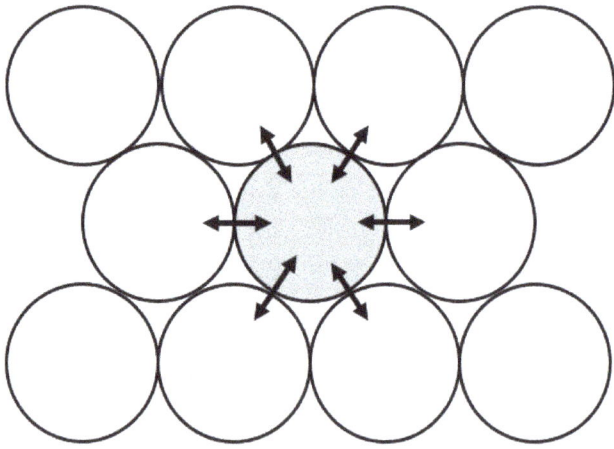

6 adjacent atoms in same layer

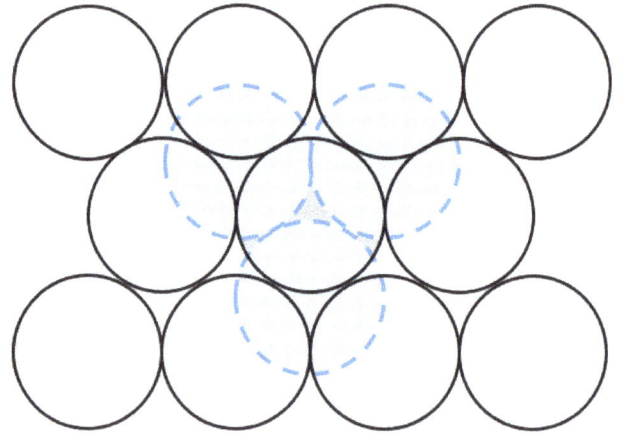

3 adjacent atoms in layer above (and below)

Using modern electron microscope techniques, it is now actually possible to "see" the atoms at the surface of a crystal structure. This is absolutely extraordinary. When I first studied chemistry in high school, no one imagined that instruments would ever be developed that could allow us to "see" individual atoms.

In fact, atoms are not quite rigid spheres, like cannon balls. The atom's outermost electron orbitals do not have a sharp boundary. When we view a scanning tunnelling microscope image (like the image shown here of the surface of copper), the atoms look like fuzzy balls (perhaps like cannon balls might look to someone who needs glasses). Nonetheless, considering atoms like rigid spheres simulates reality pretty well.

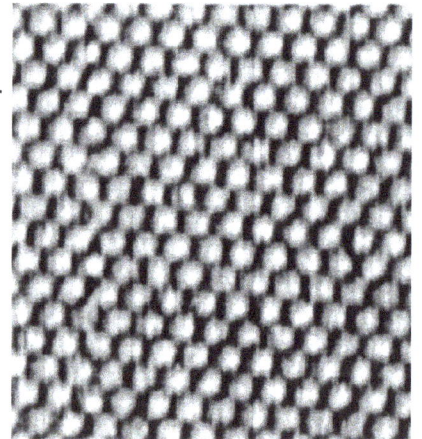

Source: Flemming Besenbacher, Scanning tunnelling microscopy studies of metal surfaces, Rep. Prog. Phys. 59 (1996) 1737–1802, p. 1788

How does this highly-ordered structure in a solid crystal compare with how atoms are arranged in the liquid? In the liquid state, the atoms still want to be surrounded (and attracted to) as many neighbours as possible, but atoms in the liquid are not as strictly ordered. Rather than being like a regiment of well-disciplined soldiers standing in formation, atoms in a liquid are like a crowd of commuters in the Tokyo subway station at rush hour.

Source: https:// resources. realestate.co.jp/ living/most-crowded-train-lines-in-tokyo-rush-hour-2017/

102

Any individual is not trapped next to their nearest neighbours, but can gradually move through the crowd (although, if the crowd is moving quickly, they can be swept along to the wrong platform). At any instant in time, an atom is surrounded by neighbouring atoms - but fewer than in a solid crystal structure. Hence, in the liquid, the energy of attraction to surrounding atoms is less. On average, a particular atom in the liquid may be surrounded by, and attracted to, (say) ten nearest neighbours. The reduced energy of attraction to surrounding atoms/molecules accounts for the "Heat of Fusion", E_m, which must be provided to melt the solid.

Let's now imagine the case where we have a crystal of solid metal surrounded by molten metal. Let's say that no heat is added or removed, so the crystal neither grows nor melts. The solid and liquid are in equilibrium. This occurs when the temperature is at the "melting point" of the solid (which is the "freezing point" of the liquid). *At the melting point, the probability that any particular atom escapes from the surface of the crystal is equal to the probability that a surrounding atom in the liquid will be "captured" onto the surface of the crystal*.

An atom within a solid crystal has less energy than it would have in the liquid (with the difference in energy being the Heat of Fusion). This situation is very similar to atoms or molecules within a liquid being in equilibrium with its vapour. Atoms can escape from the crystal surface only if they have sufficient energy to overcome the energy difference between the solid and liquid state – that is, atoms can only escape if their energy exceeds the heat of melting.

Even though atoms in a crystal are held in a relatively rigid structure, they do have thermal energy and vibrate around their assigned positions. Some atoms have more energy than others, with vibrational energy distributed exponentially among the atoms. *The fraction of atoms that have sufficient energy to escape the crystal surface is e^{-E_m/RT_m} (where E_m is the molar heat of fusion, R is the Universal Gas Constant, and T_m is the melting temperature).*

But what about atoms in the liquid that happen to find themselves immediately above the crystal surface? *Atoms in the liquid can be "captured" onto the crystal surface - but only if they happen to be in a suitable position (relative to the atoms below) to "fit" onto the crystal structure*. For most metals, as it turns out, about 40% of liquid atoms happen to be in the right place to be captured onto the crystal.

Let's consider what this might "look like" to a metal atom drifting through the liquid, immediately above the metal surface. Let's say that the metal crystal has a close-packed hexagonal crystal structure. The surface consists of identical repeating units, with each unit formed by an equilateral triangle connecting the centres of three adjacent metal atoms. I have shown these triangles with blue lines.

Take a close look at one of these units formed by three adjacent metal atoms.

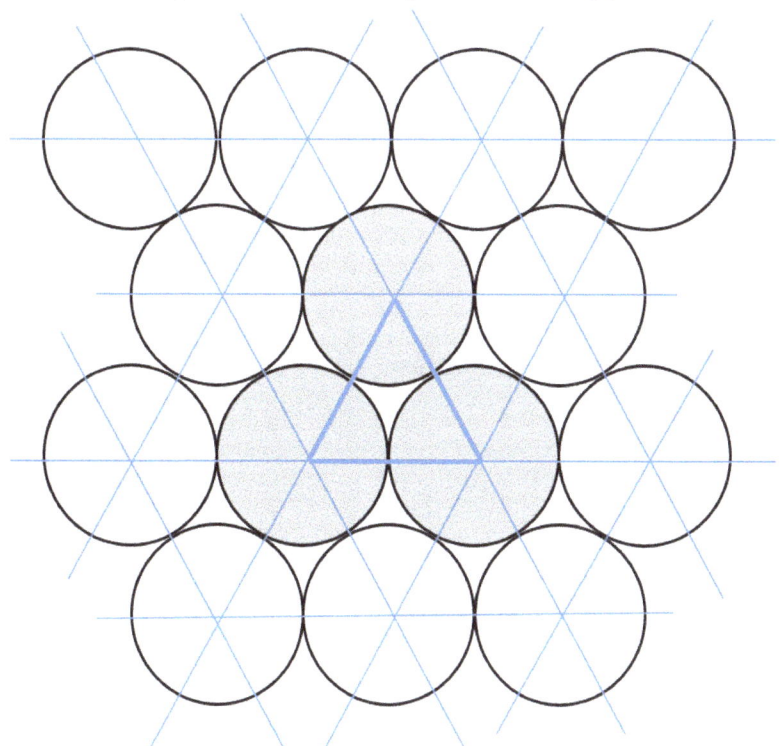

Imagine that you were a metal atom drifting in the liquid, immediately above the crystal surface. At what point on the surface would you most likely become bound to the crystal?

Point of minimum attraction

- If you drifted directly over the centre of a metal atom in the surface (at the corner of one of the triangular units), you would be attracted to - this **one** atom. This downwards attraction is less than you would feel from metal atoms in the liquid above. A metal atom landing here would be pulled back into the liquid.

- If, on the other hand, you drifted to the point at the centre of the triangle formed by three metal atoms on the surface, you would contact - and be attracted to - all **three** metal atoms. Here, you would have lower energy than in the liquid, and would become bound to the crystal surface. I have indicated this point with an arrow.

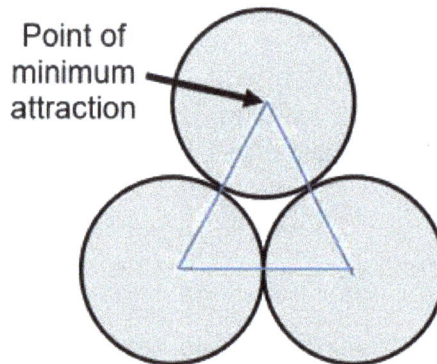

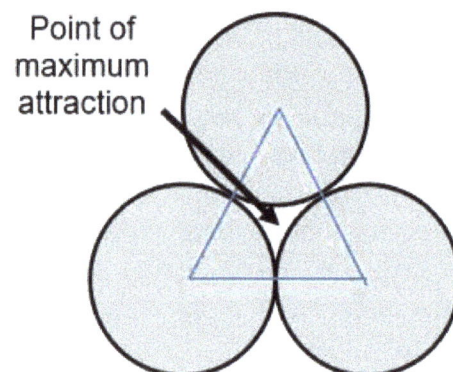

Point of maximum attraction

For a metal atom drifting randomly above the crystal surface, it would be extremely unlikely to drift to this single point on the surface exactly in-between the three atoms. However, a metal atom would likely become bound to the surface if it drifted **close** to this point. How close? Well, imagine an atom drifting within the triangular area representing each unit of the surface. As it turns out, the atom would have a 40% probability of becoming bound to the surface. This corresponds to an atom in the liquid whose centre lies within the dotted circular area that I have indicated in red [Note 3].

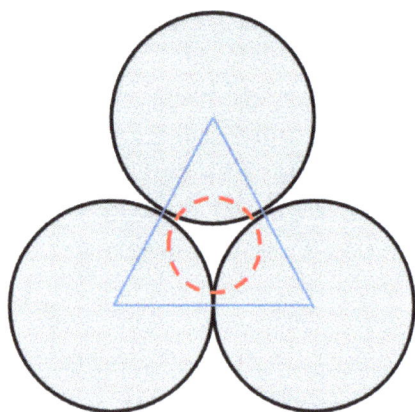

So, if a metal atom in the liquid happens to find itself above the surface within this circular area (relative to the solid surface below), it would become bound to the crystal surface. If it landed anywhere else within a triangular unit of the surface, it would be pulled back into the liquid.

At the melting temperature T_m, solid crystals are in equilibrium with molten liquid. There is a "tug-of-war" between two opposing forces, which are balanced:

- Atoms or molecules in the solid crystal have lower energy than in the liquid. **Only a fraction of atoms on the crystal surface have sufficient energy to escape**. Any atom or molecule on the crystal surface with sufficient energy can escape – regardless of its position or orientation on the crystal surface.

- **For atoms or molecules in the liquid, only a fraction are in a suitable position (or orientation) to become bound the crystal surface**. Any atom or molecule in a suitable position will be attracted and become bound to the crystal - regardless of its energy.

At the melting temperature, the probability for an atom to escape the crystal surface is equal to the probability of an atom in the liquid to be captured. We have seen that the probability for an atom having sufficient energy to escape from the crystal surface is $e^{-Em/RT}$. Let's say that an atom in the liquid directly above the crystal has probability **p** of being in a suitable position to become bound to the surface. By setting these two probabilities as equal, we get the following equation:

Equation (1) $e^{-Em/RTm} = p$

Where E_m is the heat energy absorbed in melting one mole of crystalline solid
T_m is the melting point of the crystalline solid
R is the Universal Gas Constant (8.3 Joules/mole-deg K)
p is the probability that an atom in the liquid layer above the crystal surface is in a suitable position to be bound to the crystal surface.

For the case of metals, **p** is about 0.4. Taking the logarithm of both sides of Equation (1), gives the following relation between the molar heat of melting and melting temperature:

Equation (2) $\dfrac{E_m}{T_m} = -R\ln(p)$

For metals, **p** is about 0.4, so:

$$\frac{E_m}{T_m} = \text{approximately 7}$$

This is an extraordinary result! The molar heat of fusion of metals varies in direct proportion with their melting temperature (relative to absolute zero). This is due to the nearly constant probability factor (0.4) for metal atoms to become bound to their solid crystal.

The ratio E_m/T_m is the "change in Entropy on melting", which is relatively constant for all metals. *It reflects the probability for atoms in the liquid to become bound to the solid crystal.*

How closely does the melting behaviour of metals compare to that predicted by Equation (2)?

Atoms in the first and second column of the periodic table of the elements are metals, and their electrons occupy atomic orbitals which have spherical symmetry. The melting temperature (in degrees Kelvin) and Molar Heats of Fusion are listed here and plotted on a graph. Note that all the data points lie very close to a straight line passing through the origin, corresponding to a constant value of E_m/T_m.

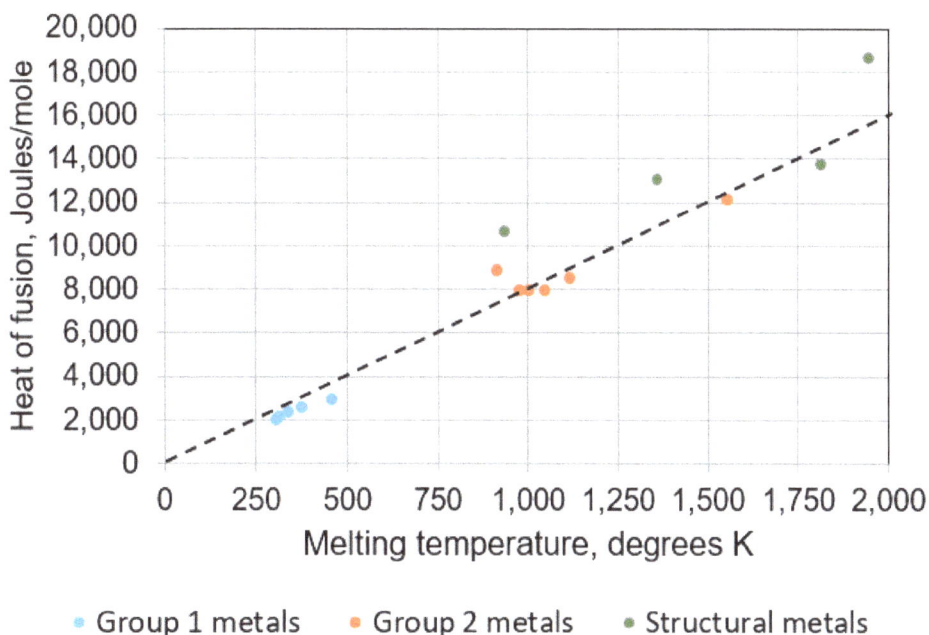

Melting points and Heat of fusion, metals

Group 1 metals   • Group 2 metals   • Structural metals

105

Most common metals that are used to make structural components and other stuff are "transition elements", whose outer-shell electron orbitals are not entirely spherical. For transition metals, the ratio E_m/T_m generally lies within the range 8-12 (corresponding to a probability **p** of 0.24-0.38 for liquid atoms to be captured onto the crystal surface).

	Metal	Melting point deg Kelvin T_m	Heat of fusion Joules/mole E_m	$\dfrac{E_m}{T_m}$
Group 1 metals	Lithium	454	3,000	6.6
	Sodium	371	2,600	7.0
	Potassium	336	2,385	7.1
	Caesium	302	2,090	6.9
	Rubidium	312	2,190	7.0
Group 2 metals	Beryllium	1,551	7,895	5.1
	Magnesium	912	8,940	9.8
	Calcium	1,115	8,540	7.7
	Strontium	1,042	8,000	7.7
	Barium	998	8,000	8.0
	Radium	973	8,000	8.0
Structural metals	Iron	1,811	13,800	7.6
	Aluminium	933	10,700	11.5
	Titanium	1,941	18,700	9.6
	Copper	1,358	13,100	9.6

Source of data: http://periodictable.com/Elements/056/data.html

Many metals have high melting temperatures, which can be attributed to two factors:
1. High attractive forces between metal atoms, so that their Molar Heat of Melting is high.
2. The spherical, or near-spherical, symmetry of their atoms.

On the other hand, many chemical compounds are comprised of molecules with elongated or complex shapes. For such molecules to become bound to a crystal surface, they must not only happen to be located at a **suitable position** relative to the underlying crystal structure, but they must also happen to have the **correct orientation** to align with the crystal. The more the shape of a molecule differs from spherical symmetry, the lower is its probability **p** of being captured onto a crystal surface, and the greater is the change in entropy on melting E_m/T_m. This lowers the melting temperature of these compounds (relative to what we would expect for its heat of fusion).

An important example is water, whose molecules are bent into a "V" shape, with a central oxygen atom bonded to a hydrogen atom on each end. There is a strong attraction between an oxygen atom on one molecule and a hydrogen atom on a neighbouring molecule. When water freezes, the molecules align within the crystal structure in a particular intricate pattern that maximises the attraction between oxygen and hydrogen atoms on adjacent molecules. For this reason, molecules of water are not as closely packed in a crystal of ice as they are in liquid water. Consequently, when water freezes, its volume increases by about 9%. This is extremely unusual. The vast majority of substances contract slightly when the liquid freezes.

The expansion of water as it freezes has profound significance for life on Earth. As water freezes, the ice floats to the surface, forming an insulating layer that prevents underlying water from freezing.

The attraction between oxygen and hydrogen atoms on adjacent water molecules is quite strong, so the molar heat of fusion of water, E_m, is quite large - more than that of lead. If water molecules were spherical in shape (like most metal atoms), we would expect water to melt at a higher temperature than lead, 327°C (600°K). Instead, water melts at only 0°C (273°K). So, if not for the non-symmetrical attraction of water molecules to each other, life (as we know it) would not exist.

We can think of the melting/freezing transition as being a tug of war between two opposing forces. Atoms/molecules "want" to be in the state of lowest energy (a solid crystal), but don't "want" to be regimented and ordered. They "prefer" to be in a more random, disordered state (a liquid), where they can move around and rotate.

At the melting point, these two opposing forces are balanced. The increase in disorder accompanying the solid-to-liquid transition is exactly balanced by the higher energy in the liquid state. If we increase the temperature even slightly, just above the melting point, the greater disorder (entropy) of the liquid "wins" and causes the solid to melt. If we lower the temperature below the melting point, the lower energy in the solid state "wins", the crystals grow and the liquid freezes.

Notes

1. Mercury is the only metal that, when pure, is liquid at room temperature. For this reason, mercury had been widely used in thermometers and other applications. Mercury has long since been replaced in such applications by other liquids due to it's very high toxicity as a neurotoxin. Nonetheless, people still often use expressions like "the mercury is rising (or falling)", which the younger generation must find quite perplexing. The low melting point of mercury (-38°C) can be attributed to its unusually low Molar Heat of Fusion (2,290 Joules/mole).

   Gallium and caesium have melting points that are just above room temperature, and will melt in your hand. Here is a website with a video showing how to cast a spoon from gallium. The spoon melts when placed into hot water. https://www.youtube.com/watch?v=cvRcUeWjBu0

2. There are two solutions to achieve the closest packing of identical spheres. Both solutions give the same packing density. One is called "hexagonal close-packing", and the other has a "face-centred cubic structure". In both stacking arrangements, the spheres occupy 74% ($\pi / (3)(2^{1/2})$) of the total volume. The structures differ only in the alignment of each third layer relative to the layers below.

3. The radius of the circular area within the triangle can be found using trigonometry. When the radius of the circle is 0.47 **r** (where **r** is the radius of the metal atoms), the area of the circle is 40% the area of the triangle connecting the centres of three atoms within the crystal structure of the solid surface.

13. Melting of alloys and solutions

Very few metals are used in pure form. To achieve the desired hardness, strength, melting point and other properties, metals are mixed with other metals or other elements. These metal mixtures are called "alloys".

Metal alloys that changed the course of history

One metal alloy which played a dominant role in development of human society was bronze, which is an alloy of copper and tin. Pure copper has very good corrosion resistance and other desirable properties but, on its own, is too soft to make good tools or weapons. About 7,000 years ago, people learned how to produce mixtures of copper and other metals (originally containing arsenic, but later predominantly tin, or also zinc).

Addition of tin increases the strength and stiffness of copper. Since mineral ores containing copper and tin are rarely found in the same geographic location, production of bronze required ores to be traded and transported long distances. Thus, one of the great advances brought about by the introduction of bronze tools was probably the development of long-distance trade, as well as the organisations and monetary systems to facilitate trade.

Bronze typically contains 88% copper and 12% tin by weight (with copper comprising 93% of the number of atoms, and with tin having a mole fraction of 7%). Addition of tin lowers the melting point of copper, making bronze alloys easier to produce and to shape into implements. This is a general feature of ***all alloys*** and ***all solutions. A mixture of metals has a lower melting point than any its component elements.*** The melting temperature of bronze (913°C) is about 170° lower than the melting temperature of pure copper.

Bronze was such an important "high-tech material" of the ancient world that the period from about 4,500 BC (the date of the earliest bronze tools discovered so far) to about 1,200 BC is called the "bronze age".

Bronze was largely replaced by iron around 1,200 BC – not because iron is stronger (early wrought iron was not as hard as bronze), but because trade in copper and tin ores was disrupted. By contrast, iron-containing minerals are widely dispersed, so each kingdom or empire had its own source of the essential ingredients to produce iron (iron ore, and charcoal or coal). The period from about 1,200 BC is called the "iron age". Some commentators argue that we now live in the "Information Age" or the "digital age", but it could also be argued that the "Iron Age" has not yet ended. Iron and steel continue to be the most important structural materials. Steel is an alloy consisting primarily of iron, but also containing up to 2% carbon by weight (comprising 7% of the number of atoms). Higher levels of carbon increase the hardness and brittleness of iron. The carbon content lowers the melting point from 1,535°C for pure iron to about 1,150°C for a typical cast iron.

Other metals are often added to steel alloys to impart particular properties. Perhaps most familiar would be stainless steel, which contains added chromium and nickel to increases corrosion resistance.

A number of alloys have been developed specifically to achieve low melting temperatures. This includes solder, which is used to provide mechanical and electrical connection of electronic components. Throughout the 20th century, most solders were lead-tin alloys (containing about 60% tin, 40% lead). These had a melting temperature of about 180°C, so they could readily be melted by a soldering iron. However, because of the toxicity of lead, lead-based solders are being phased out in many countries and replaced by alloys containing tin, silver and copper.

"Woods metal" is an alloy that was developed specifically for applications requiring even lower melting temperatures. It is widely used in sprinkler systems because of its very low melting temperature (70°C). The heat of a fire melts a plug of "Woods metal" within a sprinkler head mounted on the ceiling, allowing water to spray onto the fire below. "Woods metal" is composed of 50% bismuth, 26% lead, 13% tin, and 10% cadmium. The main constituent, bismuth, has a melting point of 271°C when pure.

Reduction of the freezing point in metal alloys and solutions

Let's consider a molten solution consisting mainly of sodium atoms, but also containing some potassium atoms. Both sodium and potassium are Group 1 metals. They have similar chemical properties, and are miscible (dissolve in each other). We would expect that attractive forces between sodium atoms and adjacent potassium atoms in solution to be about the same as the forces between neighbouring sodium atoms and between potassium atoms. We call this an "Ideal solution". Each type of atom interacts with its neighbouring atoms – whether sodium or potassium – in the same way.

But this mutual acceptance does not apply when the atoms crystallise on freezing. Potassium atoms are larger than sodium atoms, and do not readily fit into the highly-ordered structure of a sodium crystal (or at least, not without distorting the crystal structure). Similarly, sodium atoms are too small to fit within the structure of a crystal of potassium.

Here is a brief summary of the properties of these two metals:

Metal	Chemical symbol	Melting point T_m	Heat of fusion E_m
Sodium	Na	98°C (371°K)	2,600 Joules/mole-deg
Potassium	K	63°C (336°K)	2,385 Joules/mole deg

It is often convenient to refer to the main component in a solution (sodium, in this case) as the "solvent", and the minor component (potassium) as the "solute". However, we should not take the terms "solvent" and "solute" too literally. Sodium and potassium atoms can dissolve in solution in any proportion. It is somewhat arbitrary to say that "the potassium is dissolved in sodium" or that "the sodium is dissolved in potassium". The two elements are dissolved in each other.

A mixture of molten sodium and potassium behaves pretty much like an ideal solution. A sodium-potassium alloy called "NaK", consisting of 67% potassium atoms and 23% sodium atoms, is liquid at room temperature. In fact, this alloy only freezes when the temperature falls below -12°C – despite the fact that pure sodium and potassium are solid up to their respective melting points of 98° and 63°C. NaK alloy is the only metal that is liquid at room temperature, with the exception of mercury (which is now rarely used because of its high toxicity).

The unusually low melting point of NaK, combined with its high boiling point (785°C), make it ideally suited as a coolant for experimental "fast neutron" and "fusion" nuclear reactors, and as hydraulic fluid in high-temperature or high-radiation applications (such as supersonic missiles). However, while the very low melting point and high boiling point of NaK are potentially very useful, don't expect this alloy to find widespread application. Sodium, potassium and NaK alloy

are very chemically reactive, and burn or explode on contact with water or air. A two-minute video showing the reaction of NaK with water can be viewed at:
https://www.youtube.com/watch?v=10GwSbOOBqg

Let's say that we make a solution in which 80% of the atoms in solution are sodium, and 20% are potassium. Chemists would say that the "mole fraction" of sodium (symbol X_{Na}) is 0.8, and the "mole fraction of potassium" (X_K) is 0.2. The mole fractions of all components in a mixture must add up to 1.0.

Let's see what happens when we cool this solution to its freezing point. Since the solution consists mainly of sodium, we would expect that is freezing temperature would be similar to, but a bit less than, the freezing point of pure sodium.

Pure sodium freezes at its "melting point" (which is the same as its "freezing point") of 98°C. At this temperature T_m, a solid crystal of sodium metal is in equilibrium with the molten metal. As we discussed previously, the probability of a sodium atom escaping from the surface of the crystal is $e^{-Em/RTm}$ (where Em is the molar heat of fusion of sodium, and Tm is the melting/ freezing point of pure sodium). This is equal to the probability p of a sodium atom in the liquid becoming bound to the crystal surface.

Equation (1) $\qquad e^{-Em/RTm} = p$

Now consider how this differs from the situation in a sodium-potassium solution, in which 80% of the atoms (fraction X_{Na}) are sodium and 20% (fraction X_K) are potassium. An atom drifting thorough the molten metal can only become bound to the surface of a sodium crystal if its meets *two conditions*:

1. The atom must be at a suitable place above the crystal surface to become bound (of which the probability is p). This is exactly the same condition as applies to pure sodium.

2. The atom must also be sodium (of which the probability is 80%, or X_{Na}) !

So, at the freezing point of the sodium-potassium alloy solution Ta, the condition for equilibrium is:

Equation (2) $\qquad e^{-Em/RTa} = p\, X_{Na}$

> Where Em is the "heat of fusion" of the "solvent" sodium.
> R is the Universal Gas Constant
> Ta is the melting/freezing point of the sodium-potassium alloy.
> p is the probability of a sodium atom in the liquid being captured onto the crystal surface.
> X_{Na} is the "mole fraction" of sodium atoms (the "solvent") in the solution.

Equations (1) and (2) are very similar. In Equation (2), we have simply replaced the melting/freezing temperature of pure sodium Tm with the melting temperature of the sodium-potassium alloy Ta, and we have inserted an additional probability factor, the mole fraction of the potassium "solute" X_K.

We can combine and re-arrange Equations (1) and (2) to find the temperature Ta at which the alloy melts or freezes:

$$\text{Equation (3)} \qquad \text{Ta} = \text{Tm} \left[\frac{\text{Em}}{\text{Em} - R\, T_m\, \ln X_{Na}} \right]$$

Let's consider how much the freezing point of pure sodium is reduced in an alloy mixture containing potassium. Using Equation (3), we can find the "freezing point depression" $T_m - T_a$:

Equation (4)   Freezing point depression

$$\text{Tm} - \text{Ta} = \text{Tm} \left[\frac{R\, T_m\, \ln X_{Na}}{R\, T_m\, \ln X_{Na} - Em} \right]$$

Equation (4) may look complicated and intimidating, but it can be greatly simplified if the alloy is "dilute", that is, if the mixture contains mostly atoms of the solvent (sodium), and only a minor proportion of dissolved potassium. In this case, two approximations [Note 1] can be applied to simplify Equation (4), yielding a much simpler result:

Equation (5)   Freezing point depression, $\Delta T = \left[\dfrac{RT_m^2}{E_m} \right] X_{solute}$

Where **Tm** is the melting point of the pure solvent, degrees Kelvin
E_m is the molar heat of fusion of the pure solvent
X_{solute} is the mole fraction of the solute in the solution.

Equation (5) provides a rough "rule of thumb" to determine the effect of adding small amounts of alloying metals. It applies even for dilute alloy mixtures which are not "ideal solutions".

There is an even simpler "rule of thumb" to estimate how much the freezing point of a metal is reduced by dissolving small amounts of another material (a "solute"). We have seen previously that, for nearly all metals, the ratio E_m/T_m lies within the range of 7-12 Joules/mole-degree, which is roughly about the value of **R**. Inserting these values in Equation (5) gives:

Equation (6)   Freezing Point Depression $\Delta T = T_m\, X_{solute}$

So, cast iron contains up to 5% carbon by weight, which corresponds to 20% of all the atoms in the alloy. We would expect this alloy to have a melting point that is roughly 20% less than the absolute melting temperature of pure iron (1,811°K). The estimated reduction in the melting point is 0.20 X 1,811° = 362°, which gives an estimated melting temperature of 1,450°K, or 1,170°C, which is just about right.

Using Equation (3), we can determine that the melting point of a sodium alloy containing 20% potassium is about 27°C. We get virtually the same result with the simplified Equation (6). By adding 20% potassium (or indeed, *any* other atoms that dissolve in sodium), the melting point of sodium is reduced by about 70°C.

But hold on! The freezing point of this sodium-potassium alloy is less than the freezing point of potassium. Why doesn't potassium start to crystallise from the solution?

In pure molten potassium, atoms begin to crystallise at the melting point of 63°C. That's where the probability of an atom of potassium escaping from a crystal is equal to the probability of an atom being captured from the liquid onto the crystal surface. But for an atom in the liquid to be captured onto the crystal, it must strike the crystal surface at the appropriate area (which has probability **p**) *and it also must be a potassium atom (of which the probability is only 20%, or X_K)*. We can calculate the temperature at which *potassium* will crystallise from the solution using Equation (3) again. This time, though, **Tm** is the melting/freezing point of pure potassium, **Em** is the "heat of fusion" of potassium, and we must use the mole fraction of potassium X_K. When we do this, we find that potassium would not crystallise from the solution until the temperature falls to about -100°C!

To see how the melting/freezing point of the solution changes with the relative amounts of sodium and potassium, we need to plot a "phase diagram", like the one shown here. For each concentration of solution, I used Equation (3) to calculate the temperature at which solid sodium crystallises from solution (orange curve), and the temperature at which solid potassium crystallises from solution (blue curve). Whichever temperature is higher is the one at which the solution will begin to freeze. I have inserted a vertical arrow at the solution concentration with a sodium mole fraction of 80%.

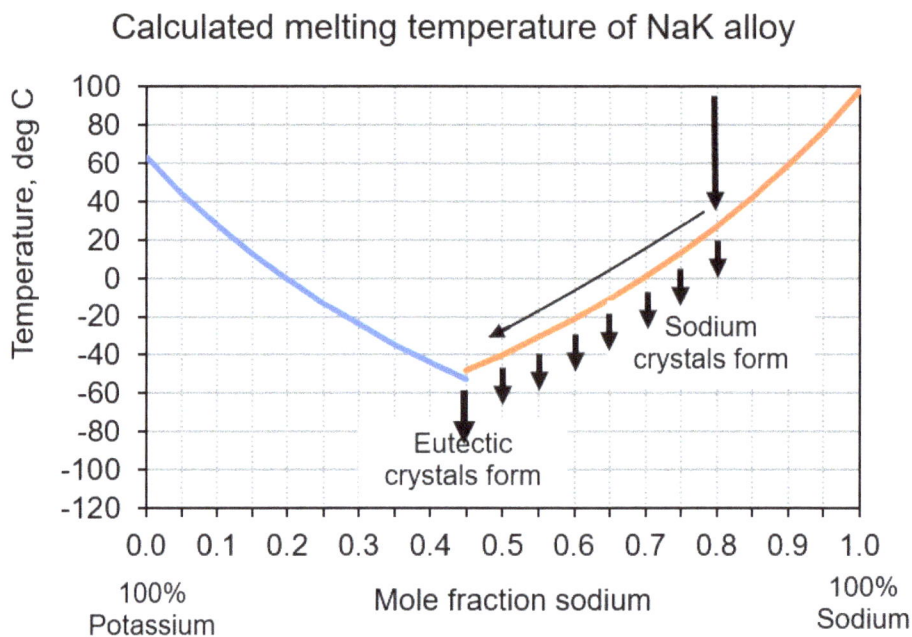

Calculated melting temperature of NaK alloy

This diagram indicates the temperature at which a sodium-potassium solution of any concentration **begins** to freeze. A solution containing 80% sodium will begin to freeze at 27°C, but as sodium crystallises from the solution, the remaining liquid becomes more concentrated in potassium. The mole fraction of sodium reduces, the mole fraction of potassium increases, and the freezing point "follows the orange curve" to the left.

As the liquid becomes more and more concentrated in potassium, its freezing point reduces – until it reaches the point where the two curves cross. This is the "eutectic concentration" at which both sodium and potassium crystallise from solution. For such a "eutectic mixture", the concentrations of sodium and potassium in the solid crystal are exactly the same as their

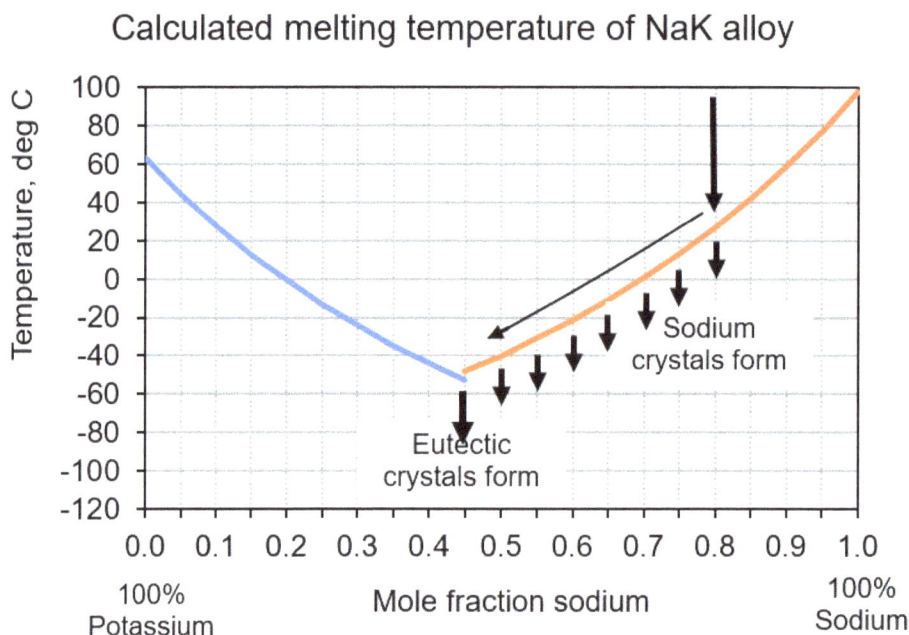

Calculated melting temperature of NaK alloy

respective concentrations in the liquid. A eutectic mixture will freeze at a single temperature – just like a pure material.

The graphs that I plotted for the freezing temperatures of sodium-potassium solutions appears similar to the phase diagram obtained by experimental measurement. However, values of melting temperatures and compositions calculated with Equation (3) are only roughly

112

indicative (my graph gives a 45% sodium mole fraction for the eutectic mixture, rather than the actual mole fraction of 77%, and a eutectic temperature of -55°C, rather than -12°). Real solutions tend to be more complicated than "ideal solutions" and often have several crystal phases. Nonetheless, Equations (3) and (4) gives a very good conceptual basis to understand the melting of solutions and eutectic mixtures.

Freezing points of solutions

The reduction in freezing point in a mixture is not limited to metal alloys. It applies to all elements and compounds that dissolve in each other.

In climates with very cold winters (including much of Europe, North America and north Asia), the freezing of water presents huge problems and challenges. A common response to this problem is to reduce the freezing point of water by adding a dissolved solute. For example, "anti-freeze" compounds are commonly added to the water coolant in car engines. Without the addition of such anti-freeze compounds, the cooling water in car engines would freeze, expand and crack the engine block. Similarly, windscreen wiper fluid would freeze solid without the addition of anti-freeze.

Virtually any compound that dissolves in water can be used as anti-freeze, but only a few compounds meet the other requirements of a good anti-freeze: being non-corrosive, relatively non-toxic and cheap. One compound that has been used is methyl alcohol (methanol). The greater is the concentration of methanol, the lower is the freezing point of the solution – but only up to a point. The lowest freezing point of a water-methanol solution is provided by a eutectic mixture, with a mole fraction of 20% water and 80% methanol. This mixture has a freezing point of -116°C, which is below the freezing point of pure methanol (-97°C). Usually, much lower concentrations of methanol (or other alcohols) are sufficient to prevent freezing in most climates.

Freezing of saltwater solutions and seawater

The freezing of water impacts in a fundamental way on the Earth's climate. Firstly, water is extremely unusual – nearly unique – in that it expands on freezing and shrinks on melting. The 9% increase in volume of freezing causes ice to float, rather than sink. In winter months, huge areas of ocean surface freeze in the polar regions, and the resulting sea-ice floats on the surface, insulating the water below and limiting the depth of the ice layer. The floating ice layer reflects most of the sunlight which strikes it (unlike seawater, which absorbs most sunlight). Reflection of sunlight by ice causes less solar heat to be absorbed, causing the Earth's surface to be cooler than it would otherwise.

Seawater is a dilute solution of salt (primarily sodium chloride) in water. In early winter in the polar regions, seawater begins to freeze when the temperature falls below -2°C. As ice crystals begin to form, the salt concentration of the remaining seawater increases, reducing its freezing point. The salt concentration can continue to increase until the solution becomes saturated at a salt concentration of 23% by weight. At this salt concentration, the freezing point of water is -21 °C. However, seawater beneath the growing ice sheets in the north and south polar regions might not reach this concentration.

A video showing how adding salt causes ice to melt and lowers its freezing point can be viewed at: https://www.youtube.com/watch?v=5m8qvQHdxuA
The video runs 5 minutes, but you can miss the first 1-1/2 minutes (in which ice cubes are ground into smaller pieces).

In the southern polar region, the continent of Antarctica is completely surrounded by ocean. In winter, as ice forms in seawater around the growing Antarctic ice shelf, the remaining seawater becomes increasingly concentrated and dense. The difference in density creates a convection current in the Southern Ocean surrounding Antarctica. Concentrated seawater formed around the periphery of Antarctica falls to the seafloor and travels northwards along the bottom of the sea, picking up nutrients from the seafloor. As this current moves north, it warms and mixes with less dense seawater, rising to the surface within a narrow belt ringing the Earth at about 50 degrees latitude. Because of the high concentration of nutrients in the upwelling stream, this area is extremely productive for the growth of crustaceans and fish, which are food sources for seals, penguins and seabirds. Consequently, islands located near this so-called "Antarctic Convergence Zone" contain some of the largest colonies of birds and marine mammals found anywhere on Earth.

Salt is also used to melt ice on roads in cold climates. This was common practice during my childhood and early adult years, when I was living in New York City. Of course, salt is highly corrosive to steel and other metals, and cars would generally rust along the "rocker panels" beneath the doors and behind the rear wheels – areas that would be sprayed with salty slush (the same areas where pebbles thrown up by the tyres chip the protective paint coating). I remember that, in my family's 1958 Oldsmobile on which I learned to drive, the bottom of the doors completely rusted out. One day, while my parents were driving along the New York State Thruway, the car hit a bump and the window of the front passenger's door literally fell through the door onto the road. Shortly thereafter, I took my first driving test, with no window on the passenger's door. It was a chilly winter day, on which you would normally expect the car windows to be rolled up. I recall that the driving examiner did a double-take when he looked down the slot in the door where the window should have been, and could see the road below. I failed that driving test, although I don't think that was due to the missing window.

When salt is added to ice or ice-water, it not only lowers its freezing point, but also causes the ice to melt. The melting of ice absorbs heat and – since salt causes ice to melt – it causes the temperature of the ice-water mixture to fall. By adding lots of salt, this temperature of the ice-salt slurry can continue to fall until it reaches a temperature of -21°C. This is the coldest temperature that could be produced by mixing salt with ice, the method used before the advent of modern refrigeration technology. Apparently, in 1724, when the German physicist Daniel Gabriel Fahrenheit proposed a temperature scale, the coldest temperature that he could produce was about -18°C, which he used to define the lower end (0°) of the Fahrenheit scale.

Notes

1. If the solution consists mainly of the sodium "solvent", the mole fraction of the potassium "solute" will be much less than one. In this case, we can use a simplifying approximation:

$$\ln X_{Na} = \ln (1 - X_K) = - X_K$$

This approximation is surprisingly accurate if the mole fraction of the "solute" X_{Na} is less than 0.3, and is even "within the right ballpark" when X_{Na} is 0.5.

Furthermore, if the solution is dilute, then the term $R\, T_m\, \ln X_{Na}$ in the denominator of Equation (4) is much less than the heat of fusion E_m and can be ignored.

About the author

Martin Gellender has always been fascinated by science. He grew up in New York City during the period after World War II, living in an environment that was economically-poor but intellectually-enriched. This was a time of awe, when people marvelled at the development of antibiotics, synthetic fabrics, skyscrapers, nuclear energy, rocket propulsion, jet aircraft and plastics. The world seemed to be on the verge of a scientific and technological revolution - and indeed, it was.

The high-rise apartment block where he lived in Lower Manhattan was basically a run-down slum, but its location was about as close as you could get to . . . the centre of the modern world. Everything was within walking distance – China Town, Little Italy, Madison Avenue, the United Nations, Greenwich Village, Fifth Avenue, Central Park, you name it.

Weekends with his family were spent on long walks to all of these places, and more. These walks often included riding a ferry across New York harbour, walking across the Williamsburg Bridge to Brooklyn, visiting the Hayden Planetarium or Museum of Natural History, Gilbert Hall of Science, browsing through the largest bookshop in the world (Barnes and Noble), touring US Navy ships when they berthed in New York, going to the observation deck of the Empire State Building (then, the tallest building in the world) or browsing through numerous shops selling "army surplus equipment" (everything from generators to bomb sights) and electronic components. These all-day walks often finished with a visit to Marty's uncle and aunt's house. There, discussions with his Uncle Sol, father and older brother often revolved around scientific developments that were underway and the possibilities for the future. Being the youngest, Marty struggled to understand what was said, but that only added to his curiosity and sense of wonder.

In 1957, the Soviet Union launched the Sputnik Satellite, and Americans were stunned that they had fallen behind their technological rival. They responded by pouring resources into scientific education. Five years later, when Marty attended a specialised high school ("Brooklyn Tech"), the facilities (machine shops, a foundry, chemistry laboratories) were state-of-the-art then (and probably better than what you would find now in any high school in Australia).

Marty went on to graduate with a bachelor's degree in Chemistry, worked for a pharmaceutical company, and did a PhD (City University of New York). He moved to Canada, where he worked as a science writer for a chemical engineering magazine (and married an Aussie), and then moved to England to start a new chemistry magazine. In 1982, he relocated to Brisbane and spent most of his career in the Queensland Government. He played an instrumental role in negotiating an agreement between the Queensland Government and CSIRO to establish the Queensland Centre for Advanced Technology (QCAT) in Brisbane, and in setting up an Energy Information Centre. Most recently, he managed a grants program that funded companies developing energy-efficient and water-saving technology.

Marty is still trying to understand how the world works, and takes great pleasure sharing his knowledge and passion with others through the University of the Third Age, and in writing this book.

www.ingramcontent.com/pod-product-compliance
Lightning Source LLC
Chambersburg PA
CBHW041620220326
41597CB00035BA/6185